PORTALS:
A TREATISE
ON INTERNET-
DISTRIBUTED
TELEVISION

Amanda D. Lotz

Published in the United States of America by
Michigan Publishing
Manufactured in the United States of America

DOI: http://dx.doi.org/10.3998/mpub.9699689

ISBN 978-1-60785-400-5 (paper)
ISBN 978-1-60785-401-2 (e-book)

An imprint of Michigan Publishing, Maize Books serves the publishing needs of the University of Michigan community by making high-quality scholarship widely available in print and online. It represents a new model for authors seeking to share their work within and beyond the academy, offering streamlined selection, production, and distribution processes. Maize Books is intended as a complement to more formal modes of publication in a wide range of disciplinary areas.

http://www.maizebooks.org

CONTENTS

Preface   v

Introduction   1
  What Is Internet-Distributed Television?   6
    Other Types of Internet-Distributed Television 9
  Understanding Internet-Distributed Television   11

1 Theorizing the Nonlinear Distinction of
   Internet-Distributed Television   15
  Theorizing Nonlinear Television   17
  Nonlinear Television as Characteristic of the
    Publishing Model?   20
  How Does Nonlinear Curation Differ from
    Linear Scheduling?   23

2 A Model For the Production Of Culture:
   The Subscriber Model   33
  A Subscriber Model of Cultural Production   39
    General Characteristics   39
    Central Function   39
    Economic Organization   42
    Creative Professions   43
    Income   45
    Market Characteristics   47
    Key Strategies   48

Implications of Subscriber-Funded Portals   51
    In What Ways Are Subscriber-Funded Portals
        "Good" and "Bad" for Audiences?   51
    In What Ways Are Subscriber-Funded Portals
        "Good" and "Bad" for Creatives?   54
    Do Subscriber-Funded Portals Enable the
        Creation of Commercial Video
        Otherwise Impossible?   56
    How Do Portal Strategies Constitute Cultures
        and Subcultures?   57
    Conclusion   58

3 Strategies of Internet-Distributed Television:
    Vertical Integration and the Studio Portal   61
    Shifts in Funding and Competitive Strategies
        before Internet-Distributed Television   62
    Vertical Integration in Internet-Distributed Television   68

Conclusion: Looking Outside Television   79

Notes   83
Index   97

PREFACE

There is a story behind everything we write. The story behind this book is that it is an illegitimate offspring of the book I've been writing, *The Cable Revolution*. I wanted to tell that story to an audience broader than media studies academics and their students, so there were necessarily ideas and deep dives into theory that didn't fit. In the process of thinking through what services such as Netflix meant to television, I discovered deeper questions about what internet distribution meant for the medium of television and how Netflix and other services were and were not like television—from an industrial perspective—and why. I dug into business and economics literature looking for insight into the peculiar business model that is quickly becoming pervasive among internet-distributed television services. Reaching intelligible answers required a lot of research that had no place in *The Cable Revolution*.

Writing up that insight started as an article, then it was two, and then I put them together because so much repetition was required. I ended up with something too long to be an article and too short to be a book, so I'm calling it a treatise, which matches its tone and provocative aspiration. It is also a conversation that doesn't fit easily into a single field. I suspect media studies folks will find this a bit too business oriented, while, well, I'll let the business folks draw their own conclusions.

In discussing a terrain that looks different every quarter, it is also necessarily incomplete and certainly partial. I offer a series of arguments about contemporary television meant to begin conversation and theory building. The focus here derives from my curiosity and thus continues an intellectual trajectory that emphasizes scripted television. Fascinating developments are occurring alongside those I investigate that are no less important (YouTube as an advertiser-funded version of internet-distributed television; live video and video redistribution by social media logics), but they are distinct enough to require their own treatises.

Its form and publication by Maize Publishing are experiments. Claiming it as a treatise seemed appropriate to its tone and brevity. Maize Publishing is an arm of the University of Michigan Press that exists for in-between things like this. Most basically, it is a lot like self-publishing, but without the parts most tedious for authors. An editorial board reviewed the project, but the manuscript did not receive blind peer review. I conducted my own peer review of sorts, soliciting feedback from colleagues, which was very helpful in revising the final draft, but they may have pulled punches more likely leveled in blind review. Thus, I'm particularly accountable for what follows, for better or worse.

My motivations for this unconventional path are multiple. First, I don't have more to say right now, certainly not enough to flesh out into a proper book. Although the terrain for internet-distributed television may be clearer in a few years, this book can make a contribution to facilitating those conversations now. Second, the speed with which Maize Publishing could take this from a manuscript to a distributed final product suited the speed at which the topic I'm writing about changes. This is a project of the moment, and new editions and subsequent works will inevitably be needed.

Finally, I am keen for the experiment. I've spent a lot of time thinking about how digital distribution changes the creation and circulation of creative products. I usually think about this in relation to video, but my chosen medium of expression is the written word. Joining the leagues of amateur video distributors isn't in my skill set, but I am curious to experience firsthand the reduction of intermediaries that digital distribution allows and see what this form of distribution allows me to learn about my readers.

What follows is certainly different than would have been the case had I waited a handful more years to publish a fuller account, but this intervention seemed needed. It is tentative in necessary ways, but deeply engages the six years of substantive internet-distributed television available.

With deep gratitude, I thank Josh Braun, David Craig, Dan Herbert, David Hesmondhalgh, Ramon Lobato, Aswin Punathambekar, and Ethan Tussey for notes on early drafts. My sincere appreciation goes as well to the University of Michigan Media Studies Research Workshop (and Jonathan Gray) and many others who have tolerated my geeky fascination with sorting this out. Thanks to Robin Means Coleman for making the time and intellectual bandwidth necessary available for this journey. I presented bits and pieces at the Inventing the New: Innovation in Creative Enterprises Conference at Northwestern University and at the World Media Economics and Management Conference, where I received valuable feedback and suggestions. I owe a considerable debt to Annemarie Navar-Gill who was an intrepid research assistant on *The Cable Revolution* and found some desperately needed information for this book as well. I am grateful for grants from the University of Michigan ADVANCE Summer Faculty Writing program and a Pohs Research Award in support of this project. My thanks to Jason Coleman and Allison Peters at Maize Publishing for their help bringing the project into the world. Sincere thanks as well to Rob Gingerich-Jones for creating the cover image, to Wes Huffstutter for being my business studies sounding board, and to Louis C.K. for inspiration.

INTRODUCTION

There is no doubt that the last two decades have produced
unfathomable changes in US television. Although a death knell
predicting the demise of television has rung loudly for both "television" and more recently for "cable," both persist, arguably both
transformed, maybe even revolutionized.[1] It is always difficult,
even impossible, to make sense of profound industrial change as it
transpires. Yet if we delay too long, some of the richness of insight
most evident in the throes of transition will be lost. During the last
twenty years, the US television industries negotiated epic shifts in
distribution and screen technologies that have had implications
for all aspects of making and viewing television. We are not without
tools to understand this change, but we must first establish what
has transpired.

Throughout the first decade of the twenty-first century, the
general presumption was that "new media" was arriving to kill
off "old media" such as television, and this perspective was so
pervasive that it obscured what has really transpired. Media cannot be killed. The written word, sound, still pictures or moving
images and the complex of industrial formations, audience practices, and textual attributes that come to define them as particular media persist. The distribution systems used to circulate
media, however, evolve with considerable regularity, and different distribution systems possess different affordances that can
introduce wide-ranging change to the production and consumption of media.[2] Only recently have "new media," as they relate

to television, been increasingly recognized not as "media" at all, but as many different platforms, technologies, messaging systems, and practices of data gathering. Many of these distribution technologies and platforms have been most widely adopted for sharing personal communication, as opposed to the intellectual property at the core of legacy media content industries. The revolutionary impact of new media upon television has not been as a replacement medium, but as a new mechanism of distribution that allows evolution of legacy companies and the creation of a sector—maybe sectors—of internet-distributed television.

Those born before 2000 were acculturated with an experience of US television that originates from the structuring requirement of a schedule. Multiple technologies and distribution mechanisms developed in the early 2000s, so it is not internet distribution alone that changes the use of television. Digital video recorders (DVRs) enabled viewers to record programming and view it on self-determined schedules beginning in the late 1990s, while cable service providers introduced video on-demand (VOD) services in the early 2000s. But adoption of both were slow despite their expansion of the capability to self-schedule and time shift—capabilities notably earlier introduced by pre-digital technologies such as VCRs and video rental. Though DVRs and VOD services are a relevant antecedent to internet-distributed television, use of these devices was too limited to meaningfully disrupt dominant industrial practices. A schedule remained necessary and central because broadcast signals could only transmit one message at a time and cable maintained the norms of the broadcast paradigm despite technological advancement. Internet-distributed television does not have this limitation.[3] Internet distribution enables personalized delivery of content independent of a schedule, which is described here as "nonlinear."

This technological capability, or affordance, and others that likewise derive from the difference in the technological mechanisms of internet distribution upend many previous television norms that were based on a single entity sending out one show to a mass audience. These capabilities alone do not require the assessment of internet-distributed audiovisual messages as distinct from the medium of television. In many cases, these messages are still produced within industrial logics consistent with broadcast- and cable-distributed television. A "medium" derives not only from technological capabilities, but also from textual characteristics, industrial practices, audience behaviors, and cultural understanding. The matrix of these factors encourages the consideration of many types of internet-distributed video to be understood as characteristic of "television."

Such an argument relies on a definition of television that derives from approaches to studying media characteristic of what was first cultural studies, and more recently considered as media studies. In one of the earliest efforts to deal with uncertainty about television in its most recent period of reconfiguration, Lynn Spigel delimited television as characterized by "technologies, industrial formations, government policies, and practices of looking."[4] Spigel precisely encapsulated comprehensive understandings of television offered earlier by scholars such as Raymond Williams, Roger Silverstone, and John Corner.[5] The recognition that all these components constituted "television"—not just a particular technological form (box, screen) or way of watching (linear schedule, on demand) was a critical intervention for identifying the consistencies that remained despite the considerable changes also occurring during what has been described as a "post-network" or "neo-network" era, or as characteristic of "TV III."[6]

The other essential contribution to a definition of television capable of incorporating its internet distribution comes from Henry Jenkins.[7] Jenkins, drawing from Lisa Gitelman, notes media are defined on two levels, as the technology that enables communication, as well as culturally, as a set of practices—or what Gitelman terms "protocols"—that develop around the technology such as the industrial practices of making television and audiences' practices of viewing.[8]

Jenkins importantly distinguishes between *media* and *delivery systems*, or what I describe as distribution technologies here. To my thinking, television is a medium, whereas broadcast signals, cable wires, and internet protocols are all delivery systems or distribution technologies. This approach diverges from Jenkins' who reserves the distinction of "media" for forms of human communication such as the written word, visual images, and audiovisual messages, rather than categorizing television as a medium. Merging Jenkins and the cultural studies tradition voiced by Spigel allows distinction among different audiovisual message systems according to their variant industrial formations and practices of looking. For example, both film and television are audiovisual messaging systems, but they are distinct media because of their discrepant industrial formations, government policies, and practices of looking.

Nearly all the "protocols" of television—per Gitelman's distinction of the term as the "huge variety of social, economic, and material relationships" connected to using technologies—are tied to television's preliminary distribution technology of broadcasting and its particular affordances and limitations.[9] Broadcast's technological ability to send just one signal led to the protocol of organizing content into a linear schedule. The schedule necessitated entities such as networks and channels as gatekeepers to organize viewing. Many norms of viewing and industrial practices that have long been believed inherent to the medium of television are, rather, protocols particular to broadcasting and cable as distribution technologies.

The affordance of internet protocol technologies to deliver personally-selected content from an industrially curated library is the central difference introduced by this new distribution mechanism.[10] The technological affordances of internet-distributed television and the varied protocols they allow encourage an industrial operation and viewer experience that is quite different from norms developed for previous mechanisms of television distribution and extend beyond nonlinearity. For example, the final two chapters of this treatise focus on the strategies of subscriber funding and the increased ability of producers to distribute content directly to consumers as strategies enabled by internet distribution that substantially change industrial norms and audience experience.

The endeavor of developing theories suited to understanding internet-distributed television requires identifying emerging protocols, investigating their similarities and differences from those of other forms of distribution, and assessing their consequences for creative practices, texts, and audiences. To that end, this treatise begins the task of theorizing internet-distributed television by establishing its distinctions, identifying its emerging associated behaviors and related logics, and beginning to create theory that addresses its particularities. The focus here examines the context of the US industry and the case of internet distribution as it exists in the United States from 2010 to 2017. There are unquestionably aspects relevant in other national contexts and careful parsing is needed to understand a distribution technology that can be global in a way previous distribution technologies were not, but this brief, preliminary assessment does not attend to these important matters.

The different affordances of internet-distributed television that enable protocols related to its nonlinearity and user specificity introduce strategies for distribution unavailable to previous mechanisms of television distribution. Sociologist John B. Thompson describes "logics" in the context of his study of the book publishing industry as "the set of factors that determine the conditions under which individual agents and organizations (that compose media industries) participate in the field," or, more colloquially, the "conditions under which they can play the game."[11] The possibility of nonlinear access characteristic of internet-distributed television allows different logics from those available to broadcast and cable television despite similar industrial formations, and to a large degree, practices of looking. Sorting out the intertwined consistency and change of internet-distributed television is crucial to developing a sophisticated understanding of television in the twenty-first century.

What Is Internet-Distributed Television?

To be clear on terms, I take television distributed using internet protocol—a method of signal distribution that disassembles messages in packets and reassembles them—as my focus. In the early years of internet-distributed video, there was a tendency to think of only video on computers as "internet television" in a manner that may confuse this distinction. Viewing device is irrelevant to this discussion. Internet protocol distribution now commonly delivers television to living room sets, mobile devices, as well as computers. The affordances of internet distribution allow strategies and practices unavailable to broadcast and cable distribution that require reconceptualization of industrial and audience practices although important similarities persist as well.

Though still a fairly recent phenomenon, internet-distributed television has had many names. To be clear, when talking about internet-distributed television, I mean the video accessed via Netflix, Hulu, Amazon Video, HBO Now and many others I'll detail momentarily.[12] Not all the video these services offer are television; many also offer feature films, which, per the difference in film's industrial formations and practices of looking, make them "film," not "television," or some other medium because they are internet distributed. The video on demand access offered by cable services is difficult to categorize. It too relies on internet protocol technology although it is bound up in industrial practices and conventions characteristic of cable. Some of the following discussion holds true for video on demand, but as of 2017, it is mostly distinct from internet-distributed television because the practices that lead to its availability derive from arrangements based on cable's linear delivery of channels.[13]

Internet-distributed television was first called "web TV," a moniker given to experiments with internet-distributed content between 2004 and 2008 that rarely included full-length, professionally produced episodes and generally predated YouTube. Notably, predictions of web TV began in 1995, which was well before most homes even had internet access. Web TV was also the brand name of a service that enabled owners to use televisions as displays for internet access from 1996 to 2013.[14] Nearly all the endeavors of internet-distributed television in this era (2004–08) failed commercially—both a result of developing before an audience for internet-distributed content existed and not offering content or a content experience deemed valuable.[15]

Another term for internet-distributed television developed within the television industry. The term OTT, an acronym of "over the top" emerged in 2005, but was uncommon until 2010. Etymologically, OTT emerged to distinguish communication that traveled—or went "over" networks built and managed by cable and telecommunications providers that was distinct from traditional cable video service.[16] OTT particularly became common in discussions about viewers choosing to end cable subscriptions (cord cutting) to instead service their video needs via providers such as Netflix and Hulu. Synonymous with internet distributed, the jargon of OTT obscures what was simply an expansion in distribution technologies.[17]

Another common industrial acronym, SVOD (subscription video on demand), was also common in industry discourse. Juxtaposed with AVOD (advertising-supported video on demand) and TVOD (transaction video on demand, more commonly known as pay per view), SVOD foregrounds the revenue model of Netflix and several other internet-distributed services. Not only do such services rely on a different distribution technology, but they also deviate from the advertiser-supported revenue model long dominant in the United States. Both SVOD and AVOD fail to distinguish between internet-distributed services such as Netflix and the video on demand services increasingly offered as part of cable video subscriptions that become robust by 2013 and have much different industrial practices. Viewers also casually used "on demand" and "streaming" as ways to describe nonlinear viewing, although often without regard for the industrial practices that distinguish the different technologies and business practices that allow this behavior.

The transition from web TV to arcane acronyms such as OTT and SVOD illustrates early thinking about the relationship of then-coming new media and television and why conceptualization of internet-distributed television remains poorly refined. All these terms obscure the consistency of television's defining attributes regardless of the development of a new mechanism of distribution with some new capabilities. Early belief of the "internet" as a form of "new media" and narratives of technological replacement concealed the reality of what has transpired—that the most desired application for distributing video via internet protocol has been accessing legacy television content outside its linear delivery.

Building an understanding of internet-distributed television based on the definitions and characterization set forth here requires acknowledging the points of commonality and distinction among television as distributed by broadcast or cable and by internet. Like earlier technologies, internet distribution requires an entity to organize and deliver programming. I use "portal" to distinguish the crucial intermediary services that collect, curate, and distribute television programming via internet distribution. Portals, such as Netflix, SeeSo, CBS All Access, and HBO Now, are the internet equivalent of channels.[18]

Although selecting content is a key task of both channels and portals, nonlinear access distinguishes portals from their channel brethren by freeing them from the task of scheduling. Portals' primary task might be better regarded as that of curation—of curating a library of content based on the identity, vision, and strategy that drive its business model. Many different curation tactics are evident among portals—tactics derived from the revenue model, the target market, and intellectual property owned by the portal, among other factors. Curation—although largely untheorized—differs considerably from scheduling, and parallels to the rich insight available about scheduling strategies must now be created for commercial library curation.

Another notable difference between portals and channels is that portals are characterized by more than just their program content, but also by the features of their interface and the capabilities they offer their viewers. There is limited differentiation in the experience of linear channels: When you turn to a channel, there is content coming through. Changing the channel yields different content, but still the same experience—perhaps only the commercial load really distinguishes the experience. In contrast, portals have features that lead to differentiation in use, and the use experience of a single portal varies among viewers. The experience of logging in to Netflix differs from what a viewer encounters entering HBO Now so that not just the programming, but also viewers' experience distinguishes portals to make portal features part of product differentiation. Some other features distinguishing portals as products include the strategy used to organize content, whether the last viewed content automatically plays, and the particular sophistication of the search and recommendation functions. Optimizing experience replaces linear scheduling strategies such as lead-ins, hammocking, etc. as mechanisms for manipulating viewer behavior. Moreover, portals offer mass customization that leads different viewers to have different experiences of a single portal. Portals are able to tailor promotional messaging and recommendations to particular subscribers rather than the mass messaging characteristic of linear distribution.

Other Types of Internet-Distributed Television

Significant variation exists among internet-distributed television endeavors. So much so that some skirt the edges of categorization as television in terms of their industrial formations and practices of looking. Although it remains early days for internet-distributed television, mounting evidence exists to support an argument that multiple video-based industries have emerged. This treatise—following the designations of television set forth—focuses on the distribution of long-form content most similar to that recognizable as "television." It consequently largely leaves unconsidered the parallel industry developing around what began as user- and amateur-generated content that has evolved into a separate industry. The emerging internet-distributed television industry that utilizes the dynamics of social media and is based on personalities that cultivate a community of followers—described by Cunningham and Craig as "commutainment" is equally fascinating and important, but distinct enough to require its own focus.[19] Clearly an emergent industry, it is defined by industrial formations (advertiser support), government policies, and practices of looking (short-form, integration of viewing into daily life) distinct enough from those characteristic of what has been "television" to be better understood as its own industry if not medium.

To be very clear, this is not an evaluative assessment suggesting any less importance of YouTube and other similar aggregators, but emerges from recognition that the particular industrial and viewer protocols of this internet-distributed video are so divergent and significant as to require their own theorization. Some aspects of the treatise's discussion of nonlinear television may apply to this sector of internet-distributed video industries as well, but for the most part, the discussion here instead recognizes the high costs of long-form, scripted production and the strategies of businesses built on circulating intellectual property as characteristic of the industrial practices of television as it has been institutionally and culturally understood. It should not be difficult to conceive of parallel television industries defined by variant logics. To a large degree, US broadcast and cable television were rightfully understood this way for most of cable's early existence.

The treatise also considers only minimally transaction-funded, internet-distributed television such as offered by iTunes or television sold on DVD. To date, such large-scale retail operations exist only as secondary markets—they are not creating original content.[20] Much of the treatise argues for models to explain emergent industrial practices for which existing theoretical frameworks are ill-suited. The "publishing model" of media production adequately explains the transaction of individual series or episodes.

Understanding Internet-Distributed Television

Although internet distribution, or more generally "the Internet" was predicted to bring seismic change to the US television industry for well over a decade before its implications became evident, what had not been predicted—really, could not have been predicted—was how other logics of the industry's operation would be affected. When the preliminary contours of a competitive space that included internet distribution began to emerge, there was a tendency to assume that the technology used in distribution could alone explain the disruption. But internet distribution has affordances unlike previous mechanisms of distribution. Indeed, technological affordances such as nonlinearity enable—and often require—substantial adjustments in the protocols of making and viewing television. These shifts are better explored in relation to their specific dimensions and the protocols they replace, rather than simply attributing them to technology. The protocols enabled by nonlinear distribution produced extensive and wide-ranging disruption of the norms that had developed for broadcast and cable distribution. It is difficult to extricate changing distribution mechanisms, preliminary revenue models, and the shifting strategies evident in US television beginning in 2010—although it is important to try; otherwise much more disruption is accorded directly to technological change than due.

Of course, internet-distributed television existed before 2010, but this year marks a significant turning point because of developments that year that made internet distribution technology more usable. The intermediary of the portal for long-form, industry-produced content emerged—in the United States—in 2010 with the surge in attention to Netflix streaming, launch of HBO Go, and expansion in robustness of Hulu.[21] The nonlinear convenience of these services made clear how television distributed by internet protocol could rival and surpass the experience of broadcast- or cable-distributed television, and countered the experience of internet-distributed video to that point as slow loading and pixelated. Moreover, the portals offered full-length, professionally produced content that was highly desired. The introduction of tablet technology in 2010 also helped bring into view what was still a coming norm of the fluid movement of "television" among an array of screens, including those previously conceived as foremost for "computing." Although the smartphone became the most important mobile screen in time, the emergence of the tablet helped shift paradigms of screen use that imagined television as bound to a television or computer. Significant attention focused on distinguishing television based on what screen was used in these early years. But internet-distributed television was soon available on the full array of screens—including the traditional living room screen—so that screen became far less important for most analyses than distribution technology.

Uncertainty about the boundaries of television that derived from its appearance on new screens occurred alongside other profound shifts in its industrial practices. A widely perceived "crisis" in the television industry and sense of demise in the early 2000s developed because the logics that had governed television and the strategies commonly used proved decreasingly effective and led to a sense of failure according to traditional benchmarks. For example, casual evaluations of the industry emphasized the severe declines in the size of the audience viewing programs live,[22] decreased viewing of returning series,[23] or forecast demise based on stagnant commitments in the upfront advertising market in 2014 and 2015.[24]

Although these were crucial metrics of the industry in an era in which broadcast networks relied entirely on advertising, they revealed only part of the story for businesses decreasingly dependent upon advertising revenue and increasingly organized by different revenue models. Metrics designed to evaluate the linear schedule, assessments of success based on live viewing, and strategies built upon the constant flow from one program to the next were so entrenched in the lived experience of television that they came to seem inherent to the medium rather than as protocols of broadcasting as a distribution system.

In setting forth this preliminary understanding of internet-distributed television from the vantage point of 2017, it is clear that the matrix of industrial change includes a new distribution technology, new screen technologies, and a previously uncommon revenue model. These alterations intersect to reveal an array of industrial strategies and audience experiences with television.

Internet-distributed television's affordance of nonlinearity enabled two key transformations in television protocols. First, it allowed the adoption of a subscriber-funded revenue model that amply adjusts many industrial practices: foremost, the aim of creating content that attracts subscribers leads to programming very different than the aim of creating content that will gather a mass of advertiser-desired eyeballs. Of course, HBO, Showtime, Cinemax, Starz, and some other linear, cable-distributed services were fully subscriber supported well before the emergence of internet-distributed television, but this revenue model was such a small component of the television economy as to attract minimal attention or theorization of these cases.[25] Certainly, not all portals utilize subscriber funding, but the extent of the adoption of this revenue model—seventy-six of the nearly one hundred portals available in the United States by the end of 2015—requires extensive rethinking of established theory about television and its industrial operations in relation to subscriber funding.[26]

The second transformative practice to derive from the non-linearity of internet distribution is the extent to which it enables the creators of television content to more directly connect with audiences. Although layers of middlemen such as internet service providers and the portals remain, television business practices are changing to allow studios greater control of production and distribution of their programming (vertical integration), than even the establishment of deeply conglomerated media companies allowed. Vertical integration among studios and networks/channels developed throughout the 1990s and altered the business of US television in ways not widely understood before the advent of internet distribution. But portals—with their central task of curation rather than scheduling—further reconfigure the strategy of vertical integration. Distinctions between producers and distributors blur in an environment of internet-distributed television in ways that produce notable consequences for the creative goods these industries develop.

Before exploring these transformations and their consequences in greater detail, this book begins by placing internet-distributed television within existing literature about media industries and establishes the conversation into which Chapters Two and Three intervene. Chapter One examines the implications of the affordance of nonlinearity. It illustrates the limitations of existing theoretical categorizations in accounting for the peculiarities of internet-distributed television and makes the case for a new model of media production, which is delineated in Chapter Two.

Chapter Two focuses on how a subscriber-funded revenue model transforms many television protocols and goes so far as describing the characteristics of a "subscriber model" of media production. The transaction of viewers paying directly for access to a bundle of content deviates significantly from models of direct payment for a single good or the advertiser-support characteristic of most existing media industries. Although subscriber funding is not particular to internet-distributed television, it is currently its predominant revenue model and has been so under considered as to require extensive analysis.

Chapter Three explores the expansion in the strategy of vertical integration evident in many recent portals. It first establishes the increased reliance on vertical integration in the US television industry in the years just before the arrival of internet-distributed television. The chapter then explores the expansion of this strategy in the emergent internet-distributed television sector through the creation of what I term "studio portals" to consider the consequences of this strategy in comparison with the norms of broadcast and cable distribution.

Chapter 1

THEORIZING THE NONLINEAR DISTINCTION OF INTERNET-DISTRIBUTED TELEVISION

The key distinction of internet-distributed television from that of broadcast or cable distribution is that it does not require time-specific viewing. Although distribution technologies using internet protocol *can* also deliver content in live and/or linear fashion, this has not been its dominant application to date. It is by eliminating the necessity of time-specific viewing that internet distribution allows different logics that extricate it from many of the conventions that have been established for television industries built on linear distribution.

The full gamut of legacy television industries' production and distribution practices, particularly those of scripted series, assumes time specificity. For decades, audiences were forced to organize their viewing according to a network-mandated schedule. Almost all the conventions of television—a flow of content, program length, expectations of weekly episodes—derive from practices developed to cope with the necessity of the linear schedule. Internet distribution advances what VCR and DVR recording and DVD access began to allow so that viewers can select television viewing just as they might select a book from a library.

In addition to the way time and timeliness has defined a particular viewing experience, time also has played a key role in circumscribing legacy business practices. The business of scripted series created for broadcast and cable distribution relies on time-based windowing of content spread over a series of markets (first, second, etc.) and licensing practices that have imbedded "time" in the core economic transactions of linear television.[1] The network licensing the original airing of a series typically enjoyed exclusive access to the series for an initial period. Much of the economic exchange between studios and networks was predicated on exclusivity that could be enforced by the scarcity characteristic of linear distribution that severely limited access to programming in an era before DVDs and internet distribution.

Linear television's time constraints also supported the advertiser-reliant business model that dominated the network era. The distribution bottleneck that yielded limited availability of content aided a system built on creating entertainment to gather an audience that would then watch advertisers' messages.[2] Of course, advertiser support was not required by this distribution technology—as robust public service television marketplaces elsewhere illustrate, but it was the American experience.

Importantly, a lot of television programming cannot be easily removed from a schedule. Much of its utility or value is connected to its liveness or correspondence to rituals of daily life, and linear broadcast and cable distribution remain effective for such content.[3] Internet distribution consequently might not be expected to wholly replace previous forms of distribution. Rather the core of its disruption will be on types of television markedly improved by the nonlinear affordance of internet distribution. This leads to a focus in the following pages on scripted series as the content form for which new logics of internet distribution are most profound.

Paradoxically, the current developments that seem revolutionary—be they internet protocol technology and related affordances, strategies related to a previously peripheral revenue model, or changing viewer behaviors that emerge from these technological and industrial shifts—are not as radical a break as has been commonly presumed. Antecedents for the contemporary environment exist as far back as the circulating libraries of the 1700s and more recently in the video rental businesses of the 1980s through 2000s. Internet distribution of television and the shifts in practices that derive from its affordances allow behaviors that were peripheral in an age of analog, physical media such as time shifting, self-curation, and á la carte access to become central and industrialized practices.

The need to explain this process of shifting cultural uses of technology brings to mind Raymond Williams' distinctions of emergent, dominant, and residual technological practices, but this framework asserts too strong a notion of replacement than evident in this case.[4] Rather, at this, admittedly preliminary point, the relationship between technologies of distribution is better described by what William Uricchio describes as profound "pluriformity" although, in time, more evidence for Williams' pattern may emerge.[5]

Theorizing Nonlinear Television

In his 1989 book, *The Capitalization of Cultural Production*, French sociologist and cultural industries theorist Bernard Miége organized the logics of media production into three models: the publishing model, the flow model, and the written press model.[6] In organizing wide-ranging media industries into these models based on their general characteristics, central function, economic organization, creative professions, business models, and market characteristics, he established "logics" useful in analysis and theory building about media industries that could be broader than single industries but still narrow enough to allow useful theory development. Within Miége's tradition of Francophone cultural industries analysis, scholars identify the logics of media industries and discern patterns of behavior, as well as possibilities foreclosed by the difficulty of deviating from the established logics, to develop critical understandings of these industries and the creative possibilities they encourage.[7]

Over the last three decades, Francophone scholars of cultural industries have defined and reconceptualized "models" through which cultural industries generally can be categorized in order to make claims of the norms and operations of media sectors and to engage in theory building broader than a particular case. Miége's triumvirate of models—publishing, flow, and written press—circulated most widely. Various refinements to the models were subsequently proposed; some suggested that the publishing model might be better classified as "commodity" production and that the dynamics of written press and flow industries were like enough to be combined.[8] Others, including Patrice Flichy, have offered different conceptualizations of larger-scale "models" relative to particular practices or "logics."[9] Jean-Guy Lacroix and Gaëtan Tremblay expanded the possible categorizations by proposing the "club logic" (really a model in Miége's use of the term) in response to the conditions described here as a subscriber model of generating revenue from media goods.[10]

The tools of cultural industries research are ideally suited for investigating the emergence of internet distribution of US television, as it is clear that much of what was known of the business of television is inadequate for understanding these new distribution technologies. The affordances of internet distribution—particularly the possibility of nonlinear distribution—establish different logics that in turn allow distinctive protocols that enable an array of practices and require an expansion of Miége's tripartite organization of models.

Miége identified the "flow model" as characteristic of radio and television and what were emerging in the late 1980s as "new media."[11] The flow model is primarily distinguished by the continuous flow of the goods produced and the way these media correspondingly become integrated into patterns of daily life. The scarcity constitutive of broadcast technology's capabilities required the distribution of a daily schedule of programming and led to its operation within the flow model. The main activity of industries operating within the flow model is as program planners who schedule the flow. Flow industries create—or arrange for the creation of—the components of the schedule (shows), but Miége argues flow industries primarily produce a *schedule* rather than particular creative goods. Broadcast and cable television distribution unquestionably followed the norms of the flow model that Miége set forth. The break from linear flow that internet distribution enables thus requires a different model for understanding nonlinear television.

Internet technologies dislodge television from the characteristics of the flow model and its attendant industrial practices so much so that some forms of television—particularly scripted series increasingly engaged through self-selection from on demand access to portal libraries—require a different model of conceptualization. Miége's "publishing model" provides another option although even this does not provide a fully explanatory model. Media that are "cultural commodities composed of isolated individual works" such as films, books, and albums characterize the "publishing model."[12] The publishers' primary duty is to organize the production and reproduction of the good, typically choosing which works will be created and identifying the creative team that will produce it.

Miége's publishing model encompassed the norms of industries such as film, the recording industry, and book publishing. The publishing model is thus not about media based upon reproducing the written word, as might be implied from the moniker, but of industries built on the creation of cultural commodities that are single works that consequently require extensive marketing. Unlike the flow and written press industries that are defined by the continuity, regularity, and the ritual of consuming their goods, a series of distinct purchases characterizes the publishing model. This attribute alters the logics of media production in a way that encourages other industrial practices. In particular, a revenue model of direct transaction payment for the media good—buying a book or album—or access to view it in the case of theatrical viewing of film dominated such industries in their analog era. Media produced within the publishing model have not substantively relied upon advertiser support.

Nonlinear distribution of television requires rethinking many assumptions of television and recalibrating models and frameworks for understanding the business of television. The practices available to internet-distributed television are distinct from those of the flow and publishing models of media operation. The protocols of portals consequently warrant deeper investigation and theory building to establish frameworks that account for their distinctions and deviations from established norms.

The affordances of internet distribution allow variant logics to govern this form of television distribution, that in turn, encourage protocols and strategies distinct from broadcast and cable distribution that have extensive implications for internet-distributed television. Shifts in industry logics—even of a less substantial nature than switching between models—adjust creative and textual possibilities to such an extent that changes in cultural goods should be expected. The emerging strategies vary television's role in culture, the creative goods television industries are most likely to produce, as well as the agency and opportunities afforded to creative talent. Only a nuanced examination reveals the points of connection and differentiation among television distributed through different technologies. Exploring the changing landscape of television as not merely a technological evolution but also with attention to shifts in logics and related practices offers a road map helpful for considering the wide-ranging disruptions digitization and internet distribution have brought to other media as well.

Nonlinear Television as Characteristic of the Publishing Model?

If the nonlinearity of internet-distributed television makes the flow model a poor fit to explain its characteristics, might the publishing model be more valuable? Some evidence can be identified that suggests Miége's publishing model could inform internet-distributed television. To be clear at the outset, there are also several limitations to understanding internet-distributed television within the publishing model, particularly that it is based on transaction payment for a particular good. In most cases, *access* to a *library* of content is transacted in the exchange between portals and subscribers, which is a remarkably different exchange.

Nevertheless, thinking about television series within the publishing model is a paradigm disrupting thought experiment that enables richer conceptualization of the consequences of internet distribution. It aids in anticipating emerging business models and practices to allow scholars concerned with critical aspects of media within society to foresee the implications of changing industrial practices related to internet-distributed video for creatives, texts, and audiences. It also strengthens the case for creating a new model better attuned to the specificities of internet-distributed television and its emerging protocols.

What would television organized according to the publishing model look like? I return to this thought experiment in the treatise's conclusion, but it likely most closely follows the experience of the transaction of books.[13] The brief television-on-DVD market that emerged in the early 2000s also hinted at this experience. Of course, television series continue to be sold on DVD, but both the monetary cost and the inconvenience of acquiring a physical good make this a preferred mechanism of distribution in few cases.[14] The internet-distributed transaction market also remains comparatively small—22 percent of the digital video marketplace in 2014, and forecast to decline in share to subscriber-funded services.[15] Although perfectly suitable as an experience, the monetary cost of transaction as a revenue model has primarily made it an option only for those for whom convenience outweighs price because competing portals offer better monetary value. Notably, with the exception of Louis C.K.'s 2016 *Horace and Pete* experiment, no series has been created for a first window of transaction sale, which makes transaction an uncertain revenue model for content creation.

Per series transaction conforms well to the publishing model, but portals that offer access to a library of content have dominated the preliminary years of internet-distributed television. Unlike the simple transaction characteristic of the publishing model in which a fee is exchanged for a single good, the portals bundle access to a range of series and movies for a monthly subscription fee. It is ironic, given the attention to "unbundling" and disaggregation as implications of internet distribution in media industries such as newspapers and music, that the most pervasive form of internet-distributed television has relied on bundling films and television series. This strategy has been explored in economics as a practice of tying,[16] bundling,[17] or as a "club model" of media industry operation[18] although none of these contexts precisely reproduces the peculiarities of the transaction of internet portals. Yannis Bakos and Erik Brynjolfsson model the economics of bundling goods sold by internet to explain the value of the strategy.[19] They identify that a seller can more precisely predict how a consumer will value a collection of goods than can a seller value individual goods. The transaction of a library as opposed to single goods is consequently valuable to both sellers and viewers.

The portals dominating internet-distributed television in the United States in 2017 do not conform to the logics of either the publishing or flow models of media production. Portals are not effectively conceptualized according to the flow model because most operate without a linear schedule. Although portals still require tasks of content selection and organization—discussed here as curation—the comparative lack of constraint that allows more than a single text to be available at a time significantly adjusts the logics and strategies of portal curation from those of broadcasting and cable. Even those portals that maintain some linear, flow tendencies—Watch ESPN, Twitch, and the autoplay experience of YouTube—provide users with such an abundance of simultaneous viewing opportunities that the strategies associated with managing the scarcity of scheduling prove suboptimal.

Moreover, in contrast to the norms of the publishing model, the underlying transaction of a portal is access to *a set of goods*, not the exchange of a specific good. Portals that rely on a subscription fee for access to a library of content thus diverge significantly from the publishing model's basis in transaction sale. This substantially differentiates the operation of most portals from the logics of the publishing model as well.

How Does Nonlinear Curation Differ from Linear Scheduling?

It is difficult to extricate the affordance of nonlinearity from the protocols it enables that in turn adjust the business of television production and distribution. But the extensive implications of nonlinearity can be seen if we compare a media industry that has never been organized by linear dictates. Juxtapose the practices of an industry such as book publishing—which has no technological capacity constraint—with linear television. Such a comparison reveals how broadcaster's capacity constraint of only making one television show available at a time deeply structures the parameters of linear television. A single channel can only distribute 24 hours of programming a day. That is a significant limit to what can be "on" in any day. Thus, the channel's ability, or requirement, to select the one thing available at any time very much defines linear television.

Linear television is consequently characterized by two related attributes: capacity constraint (limited content available) and time specificity (content available at a particular time). These attributes encouraged protocols of television experience, such as the common tendency for people to "watch television," rather than a particular show, or to sit down to "see what was on." Such behaviors and expectations derived from the sense of limited availability long perceived as inherent to television. These once-common phrases about television behavior reveal how audiences understood and experienced television, but lack pertinence in a nonlinear environment.

The attributes of linear television necessitated that the guiding determination in schedule construction for advertiser-supported television be selecting the content believed likely to attract the largest audience. Decades of familiarity with this strategy have made it seem natural, but content could have been prioritized for several other reasons: it was the best developed, it brought the most underrepresented voice, or it indicated the highest aspirations for the conventions of the medium. Even once niche address became common among precisely targeted cable channels, the primary strategy remained selecting the content likely to attract the most audience members—just perhaps the most women, children, or sports fans depending on the target of the channel.

Without scarcity governing programming strategy, it becomes possible for other tactics to emerge. Nonlinear distribution eliminates time specificity and greatly reduces capacity constraint. Of course, capacity is not limitless for nonlinear providers. But rather than distribution being the source of confinement, the cost of acquiring content becomes the key limitation. Technologically, a portal could conceivably make any piece of content ever made available, but based on current business models, the cost of licensing such a vast library would be prohibitive. Given this constraint, what strategies govern portals' selection of content?

At this preliminary point, evidence of two curation tactics can be identified for portals distributing legacy-style television. One is an *audience strategy*, and the other is a *content acquisition strategy*. The audience strategy is simply that of curating content to meet the needs of a specific audience or audience taste, especially niches not well served by existing television. For example, Noggin provides a portal with programs for preschoolers; WWE Network features programming interesting to wrestling fans. Although seemingly obvious, this is markedly different from the broadcast audience strategy of creating programming likely to gather as many people as possible.

In some ways, the strategies of audience targeting—or channel branding—that have been characteristic of cable channels seem consistent here and applicable to the portal environment. In some cases clear parallels can be identified, but the nonlinearity of portals enables a deeper deployment of this strategy. Channel branding was valuable in an expanded programming environment to allow viewers to know what they could expect to be "on" a particular channel. The nonlinearity of portals allows much more depth and the development of a library—rather than schedule—to service that interest. Moreover, since the majority of portals rely on subscriber funding, the difference in revenue model requires portals to truly serve their audience niche. In practice, this is the difference in imperative of ad-supported Nickelodeon, which seeks to attract children to sell to advertisers, and the portal Noggin, which seeks a monthly subscription fee from preschoolers' parents in exchange for access to ad-free preschooler content.[20] Nickelodeon's brand announces it as a destination for those seeking children's programming, whereas Noggin must provide enough value to warrant the subscriber fee.

Assessing the value of cultural goods is complicated and requires more extensive theory building. One effort explores how a total value of cultural goods could be comprised of multiple value measures such as nonmonetary returns, market and nonmarket use value, option value, non-use values such as existence and bequest value, and instrumental values.[21] A particularly challenging aspect of theorizing value for cultural goods is that viewers have "bounded rationality," which means they do not know their own preference for cultural goods.[22] The degree to which viewers often do not know what they want to watch explains the value of libraries that bundle multiple series.

Vastness of library is thus a key attribute for portals but must be weighed against content acquisition costs.[23] An advertiser-funded portal seeks to include as much content as possible to attract the most viewers that can then be shown advertisements. In a transaction model where viewers buy or rent particular goods, a retailer also seeks to offer as vast a catalog as possible because earnings are tied to the number of goods sold.[24] The requirement that subscriber-funded services provide something viewers value enough to pay for makes it likely that the linear niche targeting strategy will not be precisely replicated in nonlinear competition. Exclusivity becomes very important; to earn monthly payment, subscriber-funded portals need to provide content viewers want to watch—rather than just something to watch.

The more complicated site for parsing this strategy is within a portal with a broader array of content, such as Netflix. Extending the concept of cable channel branding, what is the "brand" of Netflix? Netflix's library contains content targeting multiple taste groups, and it is able to effectively target these multiple tastes because it is nonlinear. Even the most loyal Netflix consumer accesses a small amount of the library and likely has little awareness of what else is available. My description of the Netflix brand based on my viewing might differ markedly from an equally devoted subscriber with different tastes who also finds her needs met by content completely separate from what I view. Netflix takes advantage of what might be considered as the positive properties of filter bubbles so that people with different tastes have very different experiences of the content available in a way that affirms their sense of the Netflix brand.

Varying from the niche strategy of several portals then, Netflix pursues a "conglomerated niche" strategy. The company services multiple audiences, but this is very different than a "mass" strategy. It does not license or develop a series with the expectation that all Netflix viewers will value it, but develops offerings with distinct segments of subscribers in mind. Such a mass customization strategy is made possible by the elimination of the time specificity and capacity constraint of linearity that prevent channels from effectively targeting multiple audiences to achieve scale.[25] The network era comparison for Netflix is not a channel, it is a conglomerate; Netflix is not like Nickelodeon, it is like Viacom.

Such a conglomerated niche strategy achieves the advantages of scale while servicing heterogeneous tastes. What are these tastes? Only Netflix executives may know. Two of the niches targeted by Amazon Video are "people who go to Comic-Con" and "people who listen to NPR (public radio)."[26] And people with kids are clearly another target. Importantly, this is not the same as a mass strategy. Netflix achieves scale that creates efficiency for its operation, but not by being one thing to all subscribers.

Importantly, it is the attributes of nonlinearity that enable Netflix and others using a conglomerated niche strategy to be different things to different people—without anyone really noticing—and make this strategy effective. Netflix's curation strategy is guided by mass customization and its ability to market and recommend to subscribers is highly individualized. In an interview with television critic Alan Sepinwall, Netflix's chief content officer, Ted Sarandos, illustrated this logic in action by explaining how Netflix does not blanket the opening screens of all subscribers with each new show because the worst outcome is having the wrong person see new content and thus produce negative word of mouth.[27] This restraint, of only marketing a show to those likely to like it, illustrates an approach wholly uncharacteristic of the mass strategies of broadcast television, but viable and effective for nonlinear distribution. Of course, Netflix is also able to do this because of the proprietary data it collects that inform it of subscriber preferences—another affordance of internet distribution.

One subgroup of Netflix programming receives a lot of buzz and attention because it is one that closely matches that of cultural and taste opinion leaders such as critics. But evidence of multiple constituencies occasionally sneaks through. In a rare moment of data sharing, Netflix CEO Reed Hastings revealed that more subscribers watched its original horror series *Hemlock Grove* in its first days on the service than viewed the much buzzed about *House of Cards*.[28] Likewise, the announcement of a multi-film deal with Adam Sandler drew stunned silence from most who faun over the programming that services Netflix's version of the "NPR" audience. Yet, in an August 2016 interview, Sarandos noted that the Sandler movies had premiered at number one in every Netflix market and that the Sandler film, *The Do-Over*, was still in the top ten in nearly all markets three months after release.[29] Such data—if valid—reveal the conglomerated niche strategy in operation.

Of course, a key component of the portal experience derives from the mechanisms available for recommending content and helping viewers find programming that matches their interests. The portal experience is constructed through what Ted Striphas identifies as "algorithmic culture."[30] Similarly, Jeremy Wade Morris identifies the emergence of "infomediaries"—the companies responsible for shaping the cultural goods audiences encounter—that play an increasingly important role in the functioning of digitally distributed media.[31] The actions of algorithms and infomediaries allow mass customization—despite Netflix's sizable scale—to provide a seemingly individualized experience on the part of viewers.

Portals derive the most value from content that is exclusive to their libraries. Exclusivity—although long a tool of television scheduling—achieves even greater strategic use in the context of subscriber-funded portals. Unlike the norms of linear television in which a channel typically only secured exclusive rights to content for a period of time, subscriber-funded portals have sought to license original series in a way that effectively forces those who desire this content to subscribe to their services. This breaks from the prevalence of price discrimination strategies used in linear television that extracted higher economic value from audiences seeking immediate or earlier access to a series. Such strategies are less useful in a technological environment that makes unauthorized content readily available and diminishes industry-created artificial scarcity.

Also relevant to this strategy is the way nonlinear portals have tools that allow them to maximize niche strategies. As information scientists Michael D. Smith and Rahul Telang explain:

> These processes—on display at Amazon and Netflix—rely on *selection* (building an integrated platform that allows consumers to access a wide variety of content) and *satisfaction* (using data, recommendation engines, and peer reviews to help customers sift through the wide selection to discover exactly the sort of products they want to consume when they want to consume them).... They can do this because shelf space and promotion capacity are no longer scare resources. The resources that are scarce in this

model, and the resources that companies have to compete for, are fundamentally different resources: consumers' attention and knowledge of their preferences.[32]

The affordance of nonlinearity adjusts viewer experience, but the affordance of enabling data collection also adjusts the strategies portals can use in developing and curating their libraries.

Entirely separate from this audience strategy for portal curation, portals use a content acquisition strategy that leverages self-owned content to manage constraint in their program budget. Since the main limitation on a portal's library is the funds it has available to license or create programs, it is not surprising that many of the most successful portals to date are those that have been able to launch with ample content provided by corporate owners that own most or all of the content on offer. CBS All Access provides the best illustration of what I explore as a "studio portal" in Chapter Three. All of the content available on CBS All Access is owned by CBS Productions. Although most of the marketing has focused on the service as a way to watch current CBS programs, the portal also includes a significant number of library titles. In distributing through a self-owned portal, CBS seeks to profit directly from this library, rather than licensing it to another service. Notably, it also has the advantage of gaining considerable insight into the value of its content through the data generated by viewers' use of the service.

Leveraging self-owned content need not exclusively determine the content of a portal. The comedy portal SeeSo not only draws much of its content from the library of its parent company NBC-Universal, but also licenses content from others to better define its brand and provide value to its subscribers. Even the ability to leverage some content from a co-owned studio provides valuable flexibility for a portal.

Of course, vertical integration is not a strategy particular to portals; it had become a crucial strategy for linear television by the early 2000s as well. It arguably takes on heightened value for library construction because without capacity constraints, a portal's value is tied to the depth of its on-brand content. Leveraging self-owned content can help a portal offer a value proposition at launch adequate to recruit subscribers, whose fees over time can finance the expansion of that library.

Nonlinear television also provides value to audiences through convenience that may alone be more significant than any content strategy. Research on why viewers subscribe to Netflix found that 82 percent do so because of the "convenience of on-demand streaming programming," 67 percent because it is "cost-effective," and 54 percent because of its "broad streaming content library." In other words, content is only the third most compelling reason for subscription. Filling out the motivations, 50 percent subscribe because of multi-device functionality, 37 percent for kids programming, and 23 percent for original programming.[33] Such data illustrate the value inherent to nonlinear distribution.

Nonlinearity also creates new challenges. The metrics of success in linear distribution were very clear: how many watched when it aired. The value of a piece of content in a portal library is not determined so quickly. Although attempts to assess Netflix's original programs have sought counts of views within the first month of availability, measurements based on immediacy offer only a partial assessment of content value. It has long been the case that television content operates on what economist Richard Caves describes as the "ars longa" property, or a long arc of revenue return that makes television revenues durable.[34] The tremendous earnings of television series—both in the linear-only past and now—are often achieved decades after episodes are produced. Just as shows sold in syndication such as *Friends* and *Seinfeld* continue to produce considerable revenue for the studios owning them even twenty years after they ceased production, so too must the value of content held in portal libraries in perpetuity be measured over the full period of its availability.

Portals make use of the distinct logics of nonlinearity in their strategies and behaviors. Following from the focus of much of this final section on subscriber funding, the treatise now specifically explores this under-theorized revenue model and investigates the addition of a subscriber model to the flow, publishing, and written press models theorized by Miége because of its prevalence among portals in the marketplace.

Chapter 2

A MODEL FOR THE PRODUCTION OF CULTURE: THE SUBSCRIBER MODEL

Beyond the theoretical exercise of illustrating that the nonlinear affordances of internet distribution divorce portals from the flow model, recognizing the different protocols enabled by internet distribution also reveals opportunities for new strategies and ways of thinking about television's economic exchange. Nonlinear distribution and its on-demand nature require rethinking many of the strategies that have proven successful for linear television—most of which were related to scheduling. Linear distribution required synchronous viewing; it mandated timeliness as a structuring norm of television engagement.

Broadcast's technological limitations necessitated the protocol of the schedule, which was particularly well-suited for advertising because viewers had no control over the stream of content, which made them more likely to be captive for advertisers' messages. The content abundance characteristic of internet distribution and viewers' ability to self-navigate access have challenged the norms of advertising that developed for broadcast and cable distribution technologies. Legacy practices of pricing and selling advertising and measuring exposure to advertising messages have not proven effective for the very different experience of nonlinear distribution—particularly in terms of the budgets required for professional-level content. Even outside of internet-distributed video, viewer-accepted practices of advertiser messaging and valid measures for the exchange of attention remain in development and inconsistent across social media platforms. New distribution technologies such as cable-distributed, advertiser-supported, video on-demand (e.g., Comcast's Xfinity service) offer examples of access to nonlinear programming reliant in part on advertiser support, but practices specifically attuned to this environment remain preliminary.

Despite some previous cases, media have not extensively used a subscriber-funded revenue model that is not supplemented with advertising, and thus conceptualization of subscriber-funded media is limited. "Subscriptions" are most common to US media audiences in the context of magazines—which, with few exceptions, are predominantly advertiser supported despite the small subscription fee. Reliance on advertising in any amount strongly affects strategy and leads such dual-revenue media to conform mostly to logics of advertiser-supported media. US audiences also might think of a "subscription" to a cable, phone, or internet service. None of these cases is illustrative of the subscriber-funded services focused upon here. A cable subscription comes closest— it provides access to cable channels in exchange for a monthly fee. But the majority of those channels are supported in part by advertising. HBO, Showtime, Cinemax, and Starz, as purely subscriber-funded, provide the clearest precedent.

As the use of the subscriber-only revenue model has grown more prevalent among internet-distributed media, it has become necessary to theorize the specificities of subscriber-funded media as well as their use in the context of the affordances internet distribution allows. The limited use of only subscriber financing in exchange for access to an array of goods has been uncommon enough to escape detailed theorization. There are few precedents in and limited theorizing of media using a subscription-for-library-access model despite origins of this model that date to the circulating libraries of the 1700s.[1] Circulating libraries provide a precursor to portals' subscriber funding as a contemporary business model for media, as does video rental of the 1980s.[2] Subscriber funding is also a key revenue stream for emerging businesses offering streaming music access. Factors of media specificity create different dynamics in all these contexts, but nevertheless create a relevant field of examination.

Although "subscriptions" to print goods such as newspapers and magazines have been common for the last century, this transaction too is distinct from the context of portals. Miége considered these goods as produced within a "written press" model that was ultimately quite similar to the flow model because advertising provided the primary revenue stream and the regularized creation and release of these goods mirrored the scheduling function of flow media and likewise required constant production infrastructure. Moreover, subscriptions to media goods typically have been narrowly circumscribed—a subscription to a particular publication, rather than the breadth of content most portals offer. Finally, advertising remained the primary source of revenue of newspapers and magazines so that their practices of content selection are governed by strategies very different from those wholly reliant on subscriber payment.

The most significant previous work seeking to theorize subscription derives from Lacroix and Tremblay's preliminary suggestion of a "club model" in 1997. This work importantly recognized the distinctive logics of subscriber-funded media, such as cable systems that became widespread in the 1990s, but blended advertising and subscriber revenue. Moreover, their model (at least from what I can discern from translated sources) remains too broad to be extensively helpful—particularly due to its inclusion of multiple revenue models. Writing about cable and related affordances, they note, "Club logic, which develops alongside network development, offers both to those who are hooked up, that is, to those who pay a monthly subscription: programming financed by advertisements, subscriptions or on a pay-per-view basis, and products and services which consumers can ... reproduce on material supports."[3] Their work is valuable in its address of shifting distribution technologies that required refinement in previous models and begins to think through the distinctive nature of a cable subscription as a media transaction. In the US context of cable subscription at least, the cable service does not play a curatorial role comparable to subscriber-funded portals. Cable and satellite service providers participate minimally in curation or scheduling, their function is more that of the gateway utility, akin to the internet service provider in the context of portals. Cable subscriptions also mostly provide access to content reliant on advertising so that the logics of advertising govern the production of cable programming.

The clearest precursor of subscriber-funded portals is linear subscriber-funded television service as offered by HBO and Showtime. Although similar in business model, even these entities are distinct from portals because of the limitation to access created by linear transmission. As for the cases of linearly distributed HBO and Showtime, portal subscribers assess the value of subscriber-funded, internet-distributed portals based on the availability of content that matches personal interests. Subscribers might only be willing to pay for a service if it offers a value proposition considerably better than what they can achieve in an advertiser-supported environment. These services consequently embraced technologies such as multiplexing and cable on demand that expanded access beyond the linear schedule to increase their value proposition. With few exceptions, linear subscriber-funded television services were such a small part of the television ecosystem that their business model has not been extensively explored.[4]

Here, I extend the endeavors of Francophone cultural industries theorists such as Miége, Flichy, Lacroix, and Tremblay by suggesting a "subscriber model" of media industry operation. More specifically, I identify the characteristics of the subscriber model utilized by internet-distributed portals to ground it in a particular context.

Key to differentiating media operating in a subscriber model: unlike flow and even written press models, advertising does not drive these industries, and viewers buy access to a package of goods, rather than the individual goods transacted in the publishing model. As established, a "subscriber model" is not particular to internet distribution. Such a model could have been created to explain the logics and strategies evident of HBO in an era in which it remained tied to a linear schedule, and this model building consequently contributes to understanding subscriber-supported, linear television such as HBO and Showtime. Likewise, although this model building relies on the situation of internet-distributed television portals, points of commonality with various subscriber-funded music services such as Spotify and Pandora can be identified. Deeper exploration of a model for subscriber-funded entities is required now that it has emerged as the preponderant model of this preliminary stage of internet-distributed portals.

Beyond its revenue stream, a subscriber model differs from other media transactions by relying on bundling, which has been a common practice in media industries. Often the bundling characteristic of media consumption in a pre-digital age—the aggregation of articles in a newspaper or magazine, the collection of songs to make an album—emerged from the economics of manufacturing and distributing a physical good. When a physical form was required for distribution, economies of scope necessitated aggregating media to justify the manufacturing costs of the good. Internet distribution eliminates the need for a physical good and, along with viewer desire for access to disaggregated goods, has resulted in the separation of songs from albums and articles from newspapers.

The strategy of bundling access to an array of content in portals—as opposed to selling individual series as done by transaction retailers such as iTunes and Amazon—is related and yet demonstrably different. It is not that the economics of transaction functionally require bundling to make efficient sale of a good, as was the case of physical media. Rather the portal strategy of collecting goods in a library is a response to heterogeneous taste, the risk averseness of audiences to paying to try new programs, and the marketing costs of transacting single goods.

Bakos and Brynjolfsson's research, while focused on "information goods," helps explain bundling as a business practice for portals distributing television entertainment as well.[5] Their analysis seeks to explain practices emerging for goods with very low marginal costs and how bundling creates "economies of aggregation." Although their models do not fully take into account the situation of portals—particularly the dynamic of a seller that creates its own intellectual property—two points of their analysis are very relevant. The value of bundling over transaction derives from bundling's greater efficiency in predicting consumer value of a collection of goods as opposed to single goods. This greater predictability is then matched with the capacity to collect extensive data about viewer tastes that quickly generates more predictive capabilities for large bundlers that enables them to extract more value from goods than smaller bundlers. Smith and Telang address the strategy of bundles for consumers by noting the larger the bundle the more convenient for consumers, which increases their willingness to pay.[6] Bundling thus produces benefits for both sellers and consumers and is increasingly beneficial in an environment of internet distribution in which marginal cost is functionally zero.[7]

Other research has also explored the strategy of bundling in contexts in which physical goods are not a factor, such as the newspaper industry's use of subscriber funding versus pay-per-article online access,[8] bundling strategies and pricing of online magazine content,[9] or tying access to a package of channels together in cable bundles.[10] Notably, in all these cases, advertising remains a crucial revenue stream. The situation of subscriber-funded portals that exchange access to a bundle of content for a flat fee consequently remains distinctive.

Miége explains the logics of the publishing, flow, and written press models by delimiting each according to their general characteristics, central function, economic organization, creative professions, income, and market characteristics. Following that organizational logic helps identify the distinction of a subscriber model that is developed with the situation of subscriber-funded, internet-distributed portals in mind. Analysis of consequent business strategies and exploration of emerging implications follows.

A Subscriber Model of Cultural Production

General Characteristics

At its most basic, the subscriber model is characterized by a user paying a fee for access to a collection of cultural goods. The subscriber, generally either an individual or household, typically enjoys unlimited access to the collection of goods held in the library for the duration of the subscription. Media operating within this model curate a collection of cultural goods according to a strategy based on providing a particular value proposition to subscribers.

In the case of broadcast- and cable-distributed television, the limitation of the linear schedule constrained this task by enabling a channel to offer only a single piece of content at a time, but internet-distributed portals create a repository of content according to their particular curation tactic. A range of strategies differentiates the curation tactics of portals that cultivate a library of content for a taste group or conglomeration of taste groups.

The audience strategies enabled by nonlinear distribution enumerated in the introductory discussion of how library curation differs from scheduling establishes several of the general characteristics of a subscriber model. Key to subscriber-funded media is the necessity of providing content of such value that consumers will pay for it when they face a marketplace of options that includes those that do not require a subscription fee.

Central Function

The central task of media operating within the subscriber model is to curate a collection of cultural goods such that curating involves both *compiling* content and *organizing* it in a convenient and accessible manner. It is the contents of the collection and the experience of accessing it that offer the primary value propositions to subscribers. This is an important adjustment from linear subscriber-funded services that competed solely on content because its organization was confined to its ordering in a linear schedule. Although channels varied in what they offered at a particular moment, the use experience was consistent among subscriber-funded services such as HBO and Showtime and undifferentiated from those with ad-support (beyond their obvious exclusion of advertisements).

Portals accomplish the central function of collecting—or curating—cultural goods primarily either through funding content creation (original content) or licensing content from other rights holders (acquired content). Some portals license content from a wide range of sources (Netflix), whereas others use the portal as a vertical extension of self-owned creative goods. Such entities, distinguished here as "studio portals," manage the capacity constraint of content acquisition budgets through a tactic of vertical integration. Their catalogs may not be as valuable to subscribers because their offerings are not selected based on cultivating subscriber experience, but on a strategy of relying on self-owned intellectual property that diminishes the cost of operation.

The strategic opportunity of vertical integration and the reliance on a library rather than the scarce capacity of a schedule have led the portals to value ownership of rights and to maintain valued titles in perpetuity, whereas channels historically only licensed programs for an initial period. Original content consequently can continue to provide value to a portal well beyond its initial availability because there is no capacity constraint that forces elimination of some content to introduce new content. The possibility of perpetual access consequently distinguishes these subscription libraries from common practices in other media industries that have used artificial scarcity, price discrimination, and windowing to drive viewer behavior and maximize revenue.[11] Content originally created for portals thus produces long-term value for the collection and makes existing measures of success such as immediate ratings of limited value (irrespective of the irrelevance of ratings to subscriber-funded services that instead measure success by number of new subscribers and rate of cancellation by existing subscribers).

Very important to the curation strategies of these services, subscribers pay a flat fee regardless of the amount of content accessed.

Such a model responds to the variation in price sensitivity among consumers and their heterogeneous content interests. Subscriber-supported portals license rights for a period of time at a fee independent of how many viewers then access the licensed content. This makes the marginal cost for each stream of a show zero, differentiating, say, the streaming service of Netflix from its DVD by-mail service, which pays a per DVD fee in acquisition and is limited by its ability to service a single subscriber with a disk at a time.

These practices make it difficult to evaluate how valuable any single piece of the collection is to the value proposition of the service through simple measures such as how many times an episode or series has been streamed although the richness of user data that portals collect allows more sophisticated evaluations. Where advertiser-supported media have clear metrics of success based on the number of viewers attracted by a piece of content—viewers then sold to advertisers—assessing the value of each good in the library is more difficult. Some content may be accessed by a high percentage of subscribers, but not be particularly valued, while another series may be accessed by few, but that series might be so highly valued as to compel continued subscription. This is an internet-distributed corollary to Bruce Owen and Steven Wildman's broadcast television finding that "the production of mass media messages involves a trade-off between the savings from shared consumption of a common commodity and the loss of consumer satisfaction that occurs when messages are not tailored to individual or local tastes."[12] As a result, "most watched" content is not necessarily the most valuable for a service. The rich behavioral data available to portals allow for the creation of internal measures of the value of content even if these metrics and data do not circulate more broadly.

Because the experience of portal use—including search, recommendation, and user interface—also differentiates portals' value, maintaining and improving the portal product are also important functions of subscriber portals as competitive strategies of differentiation. Those who perform work related to reimagining how viewers experience content supplement the more traditional work of those who develop what subscribers can watch.

Importantly, much variation in current practices is possible that would still be characteristic of a subscriber-funded model. Portals could price according to use level rather than the all-you-can-stream norm, and license holders could seek a different remuneration model—for example, one linked to consumption. Such changes would produce considerable adjustments in portals' strategic operations.

Economic Organization

Viewers' access to a portal's curated goods is typically paid per month. Consumption within the period of subscribership is unlimited. Here, the affordance of internet distribution in enabling an individual to select content on demand marks a significant expansion in the value proposition from what linear subscriber-funded services could provide. A subscription to linear HBO often meant access to the one program HBO selected to air each hour although the service was at the vanguard of multicasting and on-demand technologies that enable viewers to derive more value from their subscription.

In exchange for payment of a monthly fee, subscribers receive log-in credentials. Contracts imply credentials are meant for only the immediate household although it should be noted that in the early years of these services, passwords were widely shared. Such behavior was tacitly permitted as services did not attempt to penalize subscribers using multiple, and sometimes simultaneous IP addresses, which may have been a strategy to cultivate eventual subscription. In some cases, fees were structured to provide simultaneous use to multiple users for additional fees. Notably, where cable subscriptions were necessarily geographically specific, portal subscriptions are not similarly tied to a particular place.[13]

In terms of their economic organization, the portals generally require extensive infrastructure and many employees tasked with redeveloping and curating the collection as well as several engaged in tasks unrelated to making or acquiring media. Curation activities include identifying and pursuing licenses for content that fits the collection, whereas others develop original content or contract for its creation. Another set of employees maintains the portal infrastructure and works on the advancement of the product. Yet another category of employees works to expand the subscriber base and provide service to existing subscribers requiring assistance.

The affordances of internet distribution that allow for gathering data about user behavior introduce more information to the tasks of curation than has been available to linear subscriber-funded services. Subscriber-funded portals thus also require employees with skill sets that enable the collection and analysis of data such as how and what subscribers view and what devices they use to do so. Such data enable more strategic curation than previously possible, and portals with larger scale achieve increased predictive abilities.[14] Notably, subscriber-funded portals use the data they collect proprietarily so that the services often know much more about viewer engagement with their content than the companies with whom they negotiate licensing deals. The availability of viewer data of much greater specificity than characteristic of previous television distribution technologies and the lack of shared, nonproprietary data provide two notable divergences from previous industry operation.

Creative Professions

The conditions and activities of creative professionals are much like those of intermediaries for linear and ad-supported television. As true of development executives at legacy networks and channels, the subscriber-funded portals require creative professionals able to curate the collection by identifying valuable licenses and to develop original content well-suited to curatorial aims. The tasks of portal curation and content development are distinct from content creation, just as broadcast network executives do not centrally participate in series production.

The creative professionals who make original content for portals follow the norms of the publishing model in a manner largely consistent with content creation for other types of television although this is a business-to-business application of the publishing model. Those commissioned to create series are employed on a series-specific basis—although some key talent, for example, actor Adam Sandler, was contracted for multiple films by Netflix in a deal that generated substantial press attention. At this point, no creative talent is exclusively tied to a portal per a model akin to the studio system although nothing precludes this.

Portals' exchange of library access rather than a single scheduled piece of content does necessitate some adjustments in established norms for contracting with and remunerating talent. Unlike previous contracts with creatives, the value of ongoing residual earnings is most limited in cases in which portals purchase series for global distribution and hold licenses in perpetuity, as Netflix has increasingly sought. In such arrangements, the lions' share of a studio's earnings is paid upfront significantly rewriting industry norms in which a successful series returned revenue for years, even decades, as it was sold and resold in a variety of domestic and international markets. This substantial revision of long-established industry business practices warrants much critical and theoretical reconsideration, as it likely has many implications for creatives, their agency in the creative process, the risks they may take, and the content they create.

Subscriber-funded portals also require creative professionals skilled in data science and analysis to evaluate subscriber behavior data because of the much richer information available to these companies. Unlike previous distribution technologies, internet distribution of video enables portal companies to produce extensive data about subscriber behavior from which they can infer preferences, predict behavior, and discern insight valuable for curators. At this point, all such data are proprietary and little of what companies know or how they use such information circulates publicly.

Subscriber portals also employ computer and data scientists to improve the functionality and experience of the portal, including recommendation algorithms, viewer interface, and all forms of functionality. Many of these roles are far removed from the content creation typically viewed as central to media industries. For example, BAMTech, the company that emerged out of Major League Baseball's early innovation in streaming, now provides the distribution infrastructure to others such as HBO Now and WWE Network; Disney bought a substantial stake in the company in 2016. Although primarily classified as "engineering roles," these duties must also be considered as expanding the range of creative work involved, not only as components of television distribution.

Income

Profitability of subscriber-funded portals requires a careful balance of limiting the cost of content licensing and creation while maintaining enough desirable content—as determined by each subscriber—to make the subscription of adequate value. The consistent economic organization among current portals yields little differentiation among services although many alternative business models are possible and would have implications for income.

Monthly fees from subscribers generate regular—and therefore fairly predictable—revenue for services in the subscriber model, and the use of automated credit card charges or bank account deductions adds further regularity to revenue flow. Once users decide to subscribe, they maintain subscriptions until actively deciding to cancel the service. Services consequently evaluate their revenue generating success based on the number of subscribers, with attention to rates of new subscriptions as well as "churn" rates, or the percentage leaving the service in a given period.

The access to a package of goods characteristic of this model thus provides more predictability than industries characteristic of the publishing model that regularly experience either steep windfalls or great losses on particular media goods. Although acquiring and maintaining subscribers is not without challenges, providing a portfolio of products and collecting extensive data about use helps these companies understand thresholds of subscriber satisfaction and subscriber preferences that aids in managing content costs.

Greater pricing variation can be achieved by tiering pricing based on the amount of content streamed. Importantly, emergent viewing behaviors are greatly influenced by the use of a pricing structure that encourages consumption through unlimited viewing. Shifts in pricing would yield changes in viewer behavior—perhaps discouraging "trying" a wide range of goods to an extent that alters subscribers' perception of value. Home internet pricing in the United States also has encouraged consumption because it too—at least through 2017—has not been priced according to use in the manner increasingly common for cellular data, but in an "all you can use" bundle adequate for all but the heaviest of internet users.[15] A change in internet pricing structures could have significant implications for portals, especially because of the uncompetitive conditions created by the prevalence of high-speed internet monopolies in the United States. This limitation might also be ameliorated once the technical capacity and pricing of mobile internet providers become a competitive alternative to home internet service.

Again, subscriber-funded portals have unprecedented access to data that inform them of the frequency with which subscribers do not finish series and a range of viewer behavior helpful in developing pricing and curation strategies of optimal value. Portals with great scale, such as Netflix, achieve a particular advantage given the vastness of data it collects. To date, these data seem to only serve internal operating strategies, but some data likely have market value if services sought to sell information. The potential value of these data is considerable. Even if they are not sold in the manner of the revenue strategies of social media companies, meaningful data are a core product created in the operation of these industries.

Market Characteristics

It is difficult to gauge market characteristics at this preliminary point of development of portals using a subscriber-funded model. A wide range of entities exists utilizing strategies from conglomerating several niches (Netflix, HBO Now), to focusing on specific genres (SeeSo), to niche focused by audience (Noggin—preschoolers), niche focused by content (WWE Network—wrestling), and as determined by existing intellectual property (CBS All Access). The monthly cost for a service must match the value proposition of the content and portal experience although curious patterns already exist. Netflix and HBO Now are the most directly comparable of the services, but HBO Now cost subscribers roughly twice as much as a Netflix subscription until Netflix raised rates in 2016 ($8, then $10 versus $15 monthly for HBO). This difference can be explained in the broader economics of the companies as Netflix maintains a direct-to-consumer relationship and retains subscriber fees in their entirety, whereas HBO Now has followed the model of partnering with providers that was characteristic of its linear service.[16] HBO sells HBO Now through partners such as Apple, Amazon, and Verizon that maintain a share of monthly fees in return for managing the customer relationship (although likely a share less than the equal share commonly claimed by cable and satellite companies distributing the linear version).[17] Given that it maintains its linear service, HBO was compelled to enter the market at a price point comparable to that service. Although an awkward comparison because most of HBO's revenue still comes from its cable-distributed service, it is also the case that HBO is much more profitable than Netflix, mostly because it spends significantly less on content. According to 2015 data published by media analyst Matthew Ball, HBO earns a monthly profit of $3.65 per subscription, whereas Netflix earns only $.28 as a result of its high programming costs, low subscription price, and the costs of international expansion.[18]

In addition to the key role of a library of intellectual property in launching a portal, a considerably sophisticated customer interface is required, which has not conventionally been a component

of the operations of many media companies with vast IP holdings. It is notable that a company the size of HBO has chosen to
distribute HBO Now through partners rather than a self-owned,
direct-to-consumer infrastructure, especially since it is part of
a conglomerate that could likely monetize that infrastructure
through other portals such as Turner's art house film service Film-
Struck and by creating portals for other Time Warner properties.

As commonly true of emergent media, the initial period considered here of significant variation among a wide range of preliminary
competitors will eventually give way to the adoption of greater stand-
ardization among fewer competitors as varied strategies prove successful, and not. This preliminary understanding of a subscriber
model and enumeration of emerging practices of internet-distributed
portals will be refined as the market matures beyond early entrants.
This analysis may be nascent, and thus limited in predictive value, but
establishing terms, taxonomies, and characteristics is valuable as a
first—although far from final—stage of theory building.

Key Strategies

Although Miége does not include "key strategies" among his aspects
distinguishing different models for the production of culture, this
too is a valuable point of analysis. Offering exclusive content is a
key strategy for subscriber-supported services—whether linear
or nonlinear. Despite this, both Netflix and HBO have utilized a
"mixed bundling" windowed strategy by offering their exclusive
original content for transaction sale by title months after original
release on the service.[19] In some cases, they also have licensed their
original content to other distribution outlets.[20] As portal strategies
mature, the different logics of subscription and the vast capacity of
libraries may lead away from this mixed strategy as only true exclusivity might promote subscription.[21]

Because portals are not confined to the linear schedule they can
maintain content and continue to derive value as long as it remains
in their libraries. Self-owned content thus can confer long-term

value, which differs substantially from previous content distribution models built upon an initial period of exclusivity and resale through multiple markets. It also makes metrics of success built for linear television difficult to apply.

The scarcity characteristic of the affordances of past technology enhanced the viability of exclusivity as a strategy. It is unclear whether exclusivity will remain as beneficial a strategy in an era of considerable abundance—arguably even a surplus—of content. Many portals have subscription arrangements more flexible than previous services that made adding and dropping services difficult, demanded start-up or initiation fees, or required lengthy contracts for service. Flexible subscription terms encourage viewer trial although may diminish the value of exclusivity. If services limit flexibility in subscription—such as through requiring or incentivizing long-term contracts—in this abundant content environment in which piracy too remains available, the tool of exclusivity may encourage illegal access rather than subscription. Though the lower price point of many portals and the convenience they offer also might drive viewers who have opted for unauthorized access into paying for access.

Vertical integration is also a key strategy of subscriber-funded portals. This strategy is discussed in depth in Chapter Three. Table 1 charts the basic features of the subscriber model in accord with Miége's format in *The Capitalization of Cultural Production*, pp. 146–47.

Table 1 Principal Logics Underlying the Subscriber Model of Internet Portals

General Characteristics

Curates a collection of cultural goods.

Individual/household purchases access to collection of goods and enjoys unlimited consumption.

Distinctive strategies for both broad and narrow collections.

(Continued)

Table 1 (Continued)

Central Function

The portal curates, purchasing rights, or creating content that can be accessed at will.

Portal maximizes value of limited content budget, sometimes by leveraging library of self-owned intellectual property as in the studio portal.

Maintains and improves viewer experience of portal technology.

Economic Organization

Large infrastructure of employees maintains the curation, licensing, infrastructure, and customer acquisition and service activities.

Actual content creation follows norms of publishing model in which employment is irregular and linked to the creation of a specific good although alternative logics are feasible (studio system).

Creator remuneration for initial creation and less linked to metric of performance; limited use of residual payments.

License fees currently paid for unlimited viewing for a period of time.

Portals aim to provide the least content that maintains maximum subscribers.

Creative Professions

Data analysts make sense of subscriber behavior to inform development/acquisition strategies.

Computer/data scientists improve recommendation algorithms, user experience interface, features, and functionality.

Development and acquisition teams oversee commissions and licenses.

Creative talent (externally contracted) makes content.

Income

Consistent; linked to number of subscribers rather than consumption of particular pieces of content or quantity of consumption.

Benefits from economies of scale and near zero marginal cost.

More variation in pricing strategies possible than currently common in market.

Market Characteristics

Varied based on general or niche aim.
Varied based on self-owned intellectual property.

Key Strategies

Goods bundled into a library.
Exclusivity.
Vertical integration.

Implications of Subscriber-Funded Portals

Claiming implications of subscriber-funded portals for viewers and creatives at such an incipient stage is difficult. But just as delineating a preliminary subscriber model aids in organizing conversation about and analysis of this rapidly evolving market, so too can establishing areas of critical focus provide guidance as more evidence becomes available. Asserting grounded claims on any of the following key critical questions is difficult given the paucity of evidence, but the following are important questions for critical assessment of internet-distributed television in its pervasive subscriber-funded deployment—both in comparing the implications of various strategies within this distribution technology and in comparison with the experience and related theories developed over decades of linear broadcast and cable distribution.

In What Ways Are Subscriber-Funded Portals "Good" and "Bad" for Audiences?

The knee-jerk response to the emergence of subscriber-funded portals has been to assume such outlets as less democratic and of particular disadvantage to those with low incomes. This is an important consideration for cultural critics to raise but should be pursued with nuance. Such assumptions disregard the way "free TV" has never been free—because advertising costs are embedded in products—and also does not account for how the advertiser-supported broadcast and cable industries have grown dependent on subscriber funding as well (explored in depth in Chapter Three). Moreover, assumptions that access to cable-distributed, "pay TV" is best predicted by income level have not borne out; rather a more complicated matrix of factors explains the roughly 18 percent of homes that did not access cable before internet-distributed services became an option. Research by the National Association of Broadcasters found that inability to pay only explained the lack of access for six of the twenty million homes without cable.[22]

By far the most costly aspect of these portals is the internet service required to use them. A line of argument can be made that internet service and the monthly fee for one or a handful of these services can still be obtained at a lower price than basic cable service. All of this is to say that claiming that a subscriber fee makes this television the terrain of the affluent is a false assumption. Nevertheless, this critique would be further moderated if the United States had a more substantive public service broadcaster that provided robust "free" service and innovatively pivoted—as has the BBC—to making its content available through a portal such as the iPlayer.[23]

Subscriber-funded portals are arguably good for audiences that have been unvalued by advertisers. Advertiser support has never been democratic, but geared toward those audiences advertisers most desire to reach. Broadcast and cable television has targeted younger, whiter, and more affluent audiences because of the way advertisers buy audiences based on such blunt demographic features. Subscriber-funded services care much less about demographic features of their subscribers beyond that they are willing to pay the monthly subscription fee. Subscriber-supported services thus have the potential to create content for audiences that advertisers have been disinterested in reaching as long as enough like-interested viewers can be aggregated to support the costs of the portal's programming.

Given the limited competition in the cable service marketplace at the launch of internet-distributed television, these subscriber portals can be argued as good for viewers because they provide more choice in configuring video expenditures. The phenomenon of households leaving or never subscribing to cable—cord-cutters or cord-nevers—has drawn considerable attention from the television industry in recent years precisely because of the fear that subscribers would replace a $80 to $100 monthly cable subscription with a $10 Netflix fee. In truth, cutting cable has been a limited phenomenon because significant internet service fees are still required and the content available through portals—even combining three or four—does not quite reproduce the content of the bundle and meet the needs for many, especially households of multiple individuals. But the competition from portals has enabled "cord shaving," or reducing the tier of cable subscription, then supplementing it with an internet-delivered service. The competition from portals, as well as internet-distributed packages of channels often called "skinny bundles" (offered by Sling and Sony in 2015 and announced for 2017 by Hulu, AT&T/DirecTV, and YouTube), have encouraged many monopoly providers to allow subscribers greater variation in packages.

Of course, these gains are not without limitations. Perhaps the biggest negative feature for viewers of subscriber-funded portals is the inability to sample content from other portals and the degree to which portals use exclusivity to drive subscriptions. Although most portals have flexible subscription plans that allow viewers short-term access and cancellation without great difficulty, such practices may change. The feasibility of unauthorized access also maintains a check against services that make access to portals onerous or unreasonably expensive.

The most widely expressed concern about subscriber-funded portals ponders their additional splintering of an already fragmented viewing culture. There is no question that the lack of time specific-ity characteristic of internet distribution further shifts US television away from understandings that were developed for the medium when it featured limited competition among a handful of channels and required viewing at network specified times. The content cir-culated by US television has not been characteristic of a "mass" medium for twenty years, and subscriber-funded, nonlinear portals only ensure greater variation of patterns of viewing. Whether this is truly "bad" for viewers, in what ways, and for which viewers—remains unexplored empirically.

In What Ways Are Subscriber-Funded Portals "Good" and "Bad" for Creatives?

Just as the affordances of subscriber-funded portals allow for dif-ferent viewing experiences, they also adjust the creative experi-ence. The different metrics of success and consequent divergent goals of a subscriber-funded outlet allow for different types of con-tent to be created because it is not constrained by the parameters of collecting an advertiser-desired audience. This can be valuable to creatives who seek to tell stories that have not been deemed as viable for that advertiser-dominated marketplace.

Beyond the opening up of storytelling possibilities, it is unclear whether the storytelling commissioned by subscriber-funded por-tals is, on balance, more good than bad for creatives. Discourses circulating through the industry—and even culturally—have sug-gested overwhelmingly good experiences of creatives who have been freed from many constraints characteristic of the linear, advertiser-supported environment in which program lengths and structures were heavily regulated and creatives received a cas-cade of "notes" from studio and network executives. The linear, subscriber-funded environment of HBO inaugurated this dis-course of subscriber-funded television as a place of great freedom and support for creatives' visions.

The strategies of a subscriber-funded service differ from those that are advertiser supported because of the discrepancy in their central mandate. Subscriber-funded services pursue strategies aimed at maintaining and attracting subscribers; advertiser-supported services seek to gather as many viewers with the characteristics advertisers desire. The differences in these mandates do encourage different programming strategies, but do not require particular approaches to creative freedom. Here, it is important to distinguish between practices that evolved into norms—such as the networks' micromanagement of creatives through notes—and strategies related to the difference in the logics of advertiser support and subscriber support. Nothing precludes an advertiser-supported service from being as creatively hands off as one supported by subscription (as the situation at FX has recently suggested), or from the subscriber-funded services to also micromanage.

There may be evidence that the creative opportunities for which subscriber-funded services are lauded come at a financial cost, and it may be a high one for creatives developing shows for subscriber-funded outlets. Guild agreements have not kept pace with changing production and distribution norms. To use one example—although audiences have rejoiced at the elimination of rerun episodes as a standard scheduling practice, the eradication of this norm has been consequential to writers, producers, directors, and in some cases, actors who garnered significant residual revenue from these airings.

The subscriber-funded portals that develop original content have further revised the economics of earnings. In most cases, shows produced for Netflix will have no "backend"; so long as the portals develop international viewer bases, they will seek perpetual and international rights leaving transaction streaming or DVD sale as the only likely secondary market. The many distribution windows characteristic of pre-internet-distributed video produced residual earnings for creatives such that a single success could yield significant financial flexibility to allow subsequent experimentation. It is too soon to appreciate the consequences of how internet distribution erodes residuals or whether subsequent guild agreements will adequately adjust to their norms, but these are necessary sites of analysis.[24]

Another curious quirk of subscriber-funded portals has been their tendency to closely guard data about viewership—even from those creating the shows they distribute. It is difficult to know whether to argue this as a good or bad attribute, as it may be case and artist specific, but the current situation—and it may be a preliminary one—in which creatives have no idea about how many viewers watch their shows and are dependent on data not externally verifiable is unprecedented and likely of consequence.[25]

Do Subscriber-Funded Portals Enable the Creation of Commercial Video Otherwise Impossible?

Although series created for portals remain a new phenomenon, the more than twenty years of HBO original series production suggests that subscriber-funded outlets do produce content unlike advertiser-supported television. Importantly, television produced within the public service mandate has likewise long illustrated other possibilities than those characteristic of US television's advertiser-dominated history.

Subscriber-funded television has been a relatively small sector of the marketplace, and more evidence of the nature of the content produced by and for these subscriber-funded portals is necessary to better understand the opportunities and limitations of this revenue model for content and for creatives. Advertiser-supported television aims to create content likely to draw the audience members sought by advertisers, while subscriber-funded services seek to provide content that justifies a subscriber's monthly fee.

In addition to considering the different imperatives of various revenue models, other related factors also differentiate content. Subscriber-funded outlets can structure storytelling differently because they do not need to allow for regular commercial breaks.

Moreover, the nonlinear affordance of portals also enables greater flexibility in program length. As discussions of the length and structure of several of Netflix early original series have noted, the ability to surpass the episode lengths of linear norms does not necessarily lead to better storytelling.[26] Finally, creatives who create for portals that release full seasons of episodes can expect a different pattern of viewer consumption than those creating series distributed by weekly episodes. Although only anecdotes from creatives support this as significant for content development, in time, analysis based on examining narrative structures and strategies will reveal the extent to which programs produced for portals may differ.

How Do Portal Strategies Constitute Cultures and Subcultures?

Much remains unknown about who subscribes—although it is a mass phenomenon—why they subscribe, and how portal viewing differs as a cultural experience from previous norms.[27] Many identified the fragmentation of the audience in the late years of cable and often regarded this "breaking up" of the mass audience as a matter of concern.[28] Portals clearly enable additional sites to further exacerbate content fragmentation while also fragmenting the audience by time because their nonlinearity does not enforce synchronous viewing.

We must be wary of viewing past cultural constitution with nostalgia and carefully weigh perceptions of lost common culture against the reality of how narrow a view of the world advertiser-supported, broadcast- and cable-distributed television permitted. It is easy for those whose culture was prevalent to see a shift from this forced, shared culture as a loss, but that shared culture was alienating and foreign to many that now see themselves or their lives represented. Moreover, the international reach of several portals further destabilizes nation and geographic proximity that were reasonably assumed of the cultural role of previous television

distribution technologies and practices and likewise have significant implications for the non-proximate constitution of cultures
and subcultures.

Conclusion

From a business standpoint, the logics of the subscriber model
have proven most successful in transitioning audiences from the
norms of broadcast and cable distribution into experimenting
with internet-distributed television. Like in the first years of cable
availability, significant heterogeneous demand exists so that
viewers desire additional programming services for wide-ranging
reasons. The subscription library model and a conglomerated
niche curation strategy correspond well with this preliminary
period of internet-distributed video, just as was the case of early
cable channel bundling, in which subscription to cable service met
the disparate desires of those seeking varied content.[29] In time,
cable channels became more narrowly targeted—or specifically
branded—but that strategy was enabled by the content creation
oligopoly that leveraged a particularly desirable channel to gain
carriage of more narrowly targeted channels in negotiations with
cable and satellite providers. Portals in many ways introduce the
long sought "á la carte" cable environment in which viewers can
more precisely select the range of program services they desire.

It is too soon to take the preliminary success of subscriber-
funded portals as a referendum on portal business models, but it
is unquestionably the case that relying only on subscriber support
makes the economic relationship many times less complicated than
advertiser support. Throughout the last twenty years, advancement
of television has been consistently instigated by subscriber-funded
services: consider the early moves of HBO and Showtime into multiplexing, then on demand, then developing original series, and
most recently, as the earliest legacy companies to develop comple-
mentary internet portals. This may not be because of an inherent
advantage to subscriber-funded models that will be ongoing, but

this simpler relationship of economic exchange is better able to embrace the uncertainty of emergent practices. The complexity of coordinating new metrics and pricing for advertising—especially with advertisers using the linear legacy model as a benchmark for the new—have made advertiser support inadequate for the scale of funding needed for the budgets of professional content.

Notably, in time, advertiser-supported television has adopted the innovations subscriber-supported entities spearheaded. Just as the revenue models and value propositions of the services that have been in the market for some time have already changed repeatedly, continued evolution should be expected based on the range of program services available and their relative specialization.

Importantly, the characteristics of a subscriber model outlined here offer little insight into portals relying on advertising support. Although revenue model is just one of many industrial factors, the distinction between advertising and subscriber support leads to very different logics and allows for a range of strategies that differentiate media businesses regardless of distribution technology. It remains a preliminary moment in the establishment of internet-distributed television, and although several strategies have emerged, measures of success and evaluation remain uncertain.

The object of study explored here may be nascent and uncertain, but it is necessary to begin building frameworks for understanding changes in the television industries such as creating taxonomies of the strategies used to differentiate internet-distributed portals and identifying characteristics of a subscriber model that are not purely hypothetical, but drawn from the operation of an emergent industry. The task of critical scholarship typically demands deeper interrogation of the consequences of matters such as industry structures or governing logics than on offer here. This initial venture finds the operation of these industries too preliminary and uncertain for broad, evidence-based claims about the consequences of subscriber-funded portals, but it is necessary to contemplate the important distinctions among these portals, their logics, and strategies from those developed for linear and ad-supported television.

Nuanced delineation of emerging models is a crucial preliminary step in developing understandings of internet distribution. Indeed, in moments of transition, practices are often too short-lived to warrant deep critical exploration—what seem like norms in one moment expire before analyses can even be published. Although existing conglomerates appear likely to dominate the portal market, the implications of vertical integration for internet-distributed television are only the same as broadcast- and cable-distributed television in the most blunt and facile senses. Rather than assuming markets and capital work the same in all situations, considerable insight can be gained by understanding divergences and what accounts for them.

Chapter 3

STRATEGIES OF INTERNET-DISTRIBUTED TELEVISION: VERTICAL INTEGRATION AND THE STUDIO PORTAL

Although it featured distinctive affordances from its outset, internet-distributed television did not appear overnight. Significant shifts from long-held norms in the competitive practices of broadcast-, cable-, and satellite-delivered television developed as internet use was introduced to American homes and just before the arrival of internet-distributed television. These adjustments in revenue models and competitive strategies in what we might now distinguish as the "legacy" industry—those television industry entities that existed before internet distribution—must be acknowledged to properly identify changes that preceded internet-distributed television from those it has wrought.

The US industry increased its reliance on subscriber funding and vertical integration in the years immediately before the launch of internet-distributed television, practices that then become even more central. It is certainly the case that the nonlinear affordances of internet distribution substantially disrupt established norms through the different protocols nonlinearity allows, but a simple story of cause and effect is made more complicated upon realization that two of the most notable adjustments are not particular to internet distribution at all.

The previous chapter explored the consequences of subscriber funding. Here, the focus shifts to vertical integration. Obviously not a new phenomenon, the stakes of vertical integration nevertheless have augmented implications as the television industry incorporates a new mechanism of distribution. Many of the portals are owned by companies that also own most—or all—of the content distributed through the portal. I categorize those portals whose primary reason of being is to leverage an existing library of content as "studio portals." The strategic benefits of vertical integration also explain the shift to conglomeration and practices of media industries that date to the film industry's ownership of theaters, but it requires new attention now because of the defining role it looks to play in establishing the competitive field of internet-distributed television. Even at the incipient moment of 2016, it appears that ownership of substantial intellectual property has already become a necessity for launching a portal. Such a barrier to entry ensures that the competitive dynamic of internet-distributed television will differ little from that of linear-based industries.

Shifts in Funding and Competitive Strategies before Internet-Distributed Television

Despite a tendency to speak of the US television industry as a single, coherent entity, this "industry" has long been composed of many different industries and businesses. Multiple, distinctive sectors compose the US television industry although they have become increasingly intertwined. For much of its "network era," most sectors were distribution businesses that constructed a schedule of programming that was distributed by broadcast signal and later also by cable wire and satellite transmission.[1]

The broadcast sector alone contains multiple distinctive businesses: the broadcast network business—based on licensing content likely to attract a mass audience that is sold to advertisers; and the station ownership business, which is based on constructing a schedule of programming for a local station and selling the viewers collected by the programs to local advertisers. Most of these stations are affiliates of the networks, a business relationship that has changed considerably over the last thirty years from one of the networks compensating the stations to be affiliates, to the stations now offering payment to the network. The station and network businesses are in principle distinct although also complicatedly intertwined because of overlapping ownership whereby the networks own several of the largest and most lucrative stations (as many as they are permitted by regulation).

Similarly, the businesses of cable service providers (as well as satellite and other wire delivery companies) sell a service to subscribers enabling access to a package of channels. This cable provider business is largely separate from that of cable channels, which, like broadcast networks, license content to construct a schedule that attracts an audience the channel then sells to advertisers. Here too, some joint ownership among cable providers and cable channels exists, but the more considerable joint ownership can be found among broadcast networks and cable channels. See Table 2 for a breakdown of these separate yet related businesses.

Table 2 Varied broadcast and cable television industry sectors

---

Broadcast

    Network business (distribution; revenue from selling audiences to advertisers)

    Station business (distribution; revenue from selling local audience to advertisers)

Cable

    Cable/satellite service (distribution; sell a service to subscribers)

    Cable channels (distribution; revenue from selling audiences to advertisers and fees from cable/satellite services)

Studios (production; make and sell content and manage intellectual property)

Separate from all these television business sectors based on distribution is the business of the studios that make content and sell or license it to distributors such as networks, channels, and most recently, portals. An arrangement known in the industry as "deficit financing" served as the dominant financing practice for creating scripted programming for much of the network era, particularly after the 1971 enactment of the Financial Interest and Syndication Rules (Fin Syn) that required networks to purchase nearly all programming from studios not owned by the network.[2] To balance the considerable risk of content creation, the networks arranged to merely license programs rather than own them outright. They paid studios roughly 70 percent of production costs, and the studios funded the deficit and maintained ownership of the series. Successful series could be sold in secondary markets, such as to local stations and international buyers, which is where the most substantial revenues of content production were earned in the network era. Cable eventually emerged as another buyer, and most recently, so too have portals such as Netflix, Amazon Video, and Hulu.

Following the elimination of the Fin Syn rules in the mid-1990s, this arrangement subtly shifted as the networks, and then cable channels, vertically integrated production and distribution by having a conglomerate-owned studio produce nearly all the programs found on their schedules. Practices became more difficult to assess once contained entirely within a single corporation. From the 1990s through early 2000s, commonly owned studios and networks mostly operated as distinct divisions of conglomerates such as Disney, Viacom, and News Corp., and performance of the studio business was evaluated independently from the network. This enabled a continuation of previous norms so that studios and networks largely operated separately despite the increased efficiency their common ownership enabled. Over time, these entities became more intertwined—although often still maintained separate "studio" and "network" executive corps.

By way of illustration, the elimination of the Fin Syn rules allowed for NBC Studios to become the predominant source of

programming aired on the NBC Network although by accounting measures, the NBC Network still paid NBC Studios roughly
70 percent of a show's production costs to license a show,
and NBC Studios, which financed the 30 percent deficit, then
counted the revenue earned by selling the series in secondary
markets on its divisional balance sheet. Licensing series from
conglomerate-owned studios better ensured the pursuit of
like-interest between the studio and network although they
remained as separately evaluated entities within their conglomerate structure.[3]

Although never governed by Fin Syn, cable channels replicated the earlier broadcast practice when they began producing
scripted series in the early 2000s. Many purchased their first
original scripted series from externally owned studios because
original cable series were viewed as very risky and lacking significant secondary market value to an extent that executives at
commonly owned studios viewed them as too risky.[4] As these
programs succeeded, the cable channels followed the dominant
broadcast practice and created their own studios—the first in the
early 2000s, but many around 2010—to produce the shows cable
channels scheduled. For example, after Lionsgate, the studio that
produced *Mad Men*, reaped the windfall of its licensing revenue
and then Sony did likewise with *Breaking Bad*, AMC created AMC
Studios in 2010 to produce subsequent series in order to maintain a financial interest in revenue earned in secondary markets.
Expansion of vertical integration at this time was also evident
as these conglomerates owning US cable channels purchased or
established channels in international markets to expand economies of scale.

So then, the common ownership of production and distribution entities allowed by the elimination of the Fin Syn rules and the
outcome of a wave of mergers and acquisitions in the early 1990s
prepared a strategic pivot of industrial practice and strategy that
was initiated before competition from internet-distributed television forced further change. Although vertical integration has

commonly been understood as a competitive strategy, it also suggested the core revenue model of legacy television industries was shifting away from the historic reliance on advertising. Television production and distribution remained profitable as separate businesses, but the elimination of the Fin Syn rules and the conglomerated ownership structures that emerged throughout the 1990s enabled a diversification of revenue that decreased dependence on advertising and made intellectual property important to the network and channel businesses as well.

At the same time that reliance on intellectual property diversified revenue and adjusted strategies of the US television industry, networks and channels further modified revenue streams by expanding their reliance on subscriber funding. Most cable channels derived half their revenue from fees paid by cable service providers on a monthly per household basis that subsidized revenue received from advertising. This subscriber revenue increased during the 1990s and 2000s to become the dominant revenue stream for many cable channels. For example, FX president John Landgraf noted that when he began at FX in 2004, 55 percent of the channel's revenue came from advertising, but by 2015, advertising accounted for just 34 percent of revenue and was diminishing by a percent a year.[5] Although many continued to think of cable channels as primarily advertiser supported, this shift in dominant revenue source left the channels more accurately described as subscriber supported and advertising subsidized. More than just an economic adjustment, this shift was important for explaining changes in programming strategies during the early 2000s.[6]

This decreasing reliance on advertising was not limited to cable, but broadcasters also adopted cable's dual-revenue stream by 2010. Broadcasters began receiving an equivalent of subscriber revenue from cable providers in 2006 in the form of retransmission fees.[7] By 2010, the business model of the networks and station owners mirrored that of cable channels, and analyst Michael Nathanson notes 2014 as a "watershed moment" because revenue from cable and satellite providers overtook advertising dollars for the first time.[8] By 2015, retransmission fees accounted for $6.3 billion dollars of broadcast network revenue.[9] This was valuable to the networks that received the full amount of the fees for their owned and operated stations and often half of the retransmission fees accrued by affiliate stations.

This interregnum period—after the period of Fin Syn rules and the conglomeration of once separate television businesses, yet before meaningful internet distribution—provides necessary background for considering adjustments in business practices related to the arrival internet-distributed television and the nonlinear distribution it allows. A series of structuring microeconomic adjustments redefined the business of the US television industry from the mid-1990s through 2010 that were largely separate from the substantial disruption that internet distribution of video introduced beginning by the end of that period. Such preliminary adjustments were responses to declining audience size as viewers spread across the expanding number of channels, as well as the financial crisis of 2008, all of which occurred just before internet-distributed competitors further fragmented audiences and posed alternatives to the dominant ad-supported system. Additionally, by this point the industry was twenty years and an executive generation removed from the Fin Syn rules and its related practices, which allowed its paradigm of separation between studios and networks to fade.

Importantly then, the television conglomerates diversified their television businesses to create three revenue streams: advertising revenue from selling the audience collected by the schedule of programming offered by networks and channels, retransmission/affiliate fees from cable/satellite/telco service providers in exchange for carriage of the network/channel on multichannel systems, and revenue from licensing studio-owned intellectual property in various traditional (syndication, off-net cable, international) and emerging secondary markets (internet-distributed program services, DVD, electronic sell-through, e.g., iTunes).

Vertical Integration in Internet-Distributed Television

Beyond the expanded use of subscriber funding, the other profound adjustment in industrial operations introduced by the affordances of internet-distributed portals was the capacity to reduce some of the layers of middlemen—or intermediaries in more academic terms—between those who make television and those who watch it. Despite the pre-internet distribution integration just chronicled, US television remained structured by a series of "bundles" that significantly inflated costs for consumers, adding as much as 100 percent to annual television ecosystem revenues according to an analysis by Needham and Associates.[10] As their analysis explains, the US broadcast and cable television industry featured bundles of bundles: individual shows are bundled into channels; channel owners such as Viacom, Disney, Time Warner, and Discovery sell bundles of channels to multichannel video programming distributors (MVPDs) such as Comcast or DirecTV; MVPDs bundle channels into tiers and sell access to viewers; and MVPDs bundle these video channels with internet access, home phone, and mobile phone service.

This reliance on bundling—particularly, the inability to separate content from channels and channels from cable packages—in combination with the lack of a competitive marketplace for internet or cable service in much of the country created a peculiar dynamic for US television distribution. Before the arrival of internet distribution, the most significant noncompetitive knot derived from the exceptional leverage content creators could exert upon MVPDs. With its hundreds of channels, it seemed the US television landscape was abundantly competitive, but analysis in 2013 revealed that nine companies (Disney, Fox, Time Warner, Comcast/NBCUniversal, CBS, Viacom, Discovery, Scripps, and AMC) controlled about 90 percent of professionally produced television content in the United States—illustrating the extent of the late 1990s conglomeration.[11] The result was that each of these content companies held the rights to some channel or content that it could use to leverage rich subscriber fees from MVPDs in a practice similar to the block booking film studios used in the 1930s and 1940s. For example, if a cable service wanted to carry "essential" content such as ESPN from Disney or Nickelodeon from Viacom, these content owners would require the systems to carry several other channels on the most widely available tiers (e.g., FreeForm, Disney Jr, Toon Disney, and Disney HD; or VH1, VH1 Classic, Spike, and Logo in the case of Viacom). It was this uncompetitive dynamic that allowed the increased reliance on subscriber fees even within an "advertiser-supported" television sector.

Chart 1, derived from a *Variety* report, illustrates how the nine companies that control professionally produced content distribute it through a multiplicity of commonly owned channels.[12]

The content oligopoly exerted unchecked power over MVPDs largely out of view of consumers who simply blamed providers for escalating fees. The arrival of satellite competition in the mid-1990s offered many viewers a competitive choice between cable or

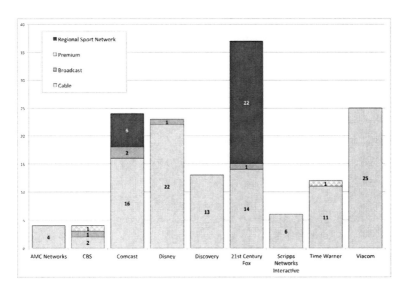

SOURCE: *VARIETY*, DRAWN FROM COMPANY REPORTS, SNL KAGAN, AND RBC CAPITAL MARKETS ESTIMATES.
EXCLUDES AMC NETWORKS' INTEREST IN BBC AMERICA, CBS' INTEREST IN POP AND THE CW, COMCAST'S INTERESTS IN FOUR CABLE CHANNELS AND TWO RSNS, TIME WARNER'S INTEREST IN THE CW, AND VIACOM'S INTEREST IN EPIX.

Chart 1 Number of Networks and Channels Owned by Company.

satellite service, but content owners forced everyone in the marketplace to provide the same value proposition—a large bundle of channels. Several cable channels began spending much more on original content in the late 1990s through early 2000s and required higher payments from MVPDs in affiliate fees in exchange. This, in combination with an explosion in the number of channels following the launch of digital cable systems in the late 1990s significantly increased viewers' cable fees. Viewers grew frustrated with service providers, but the content oligopoly left providers little power in this situation. Content creators required higher fees and bundled packaging; MVPDs passed the costs content companies demanded onto consumers.

The relationship between MVPDs and consumers was a tinderbox ready for explosion when internet-distributed television arrived. Once net neutrality rules prevented content owners from paying for preferential distribution, a new distribution dynamic became possible. MVPDs, which had quietly morphed from cable service providers to internet service providers during the first decade of the twenty-first century, benefitted the most as they often were monopoly home internet service providers. The effects of viewers shifting away from video distributed by cable—so-called cord cutting or cord shaving—were limited for providers because such consumers could be pushed into higher- and highest-priced internet service tiers, and the profit margins on internet were much better for MVPDs than video because it had no programming costs. But viewers transitioning to internet-distributed television substantially affected cable channels. Every lost subscriber meant lower fees paid by MVPDs and a smaller audience base to sell to advertisers.

The emergence of internet-distributed television could have enabled significant adjustments in the television business—and initially seemed likely to—although by 2017 it appeared that the conglomerates that dominated broadcast and cable distribution would maintain their preeminence. Importantly, a few new players—particularly Netflix and Amazon—entered the industry.

Netflix sneaked in by recognizing the opportunity of internet distribution while legacy entities were trying not to hasten disruption to broadcast and cable distribution businesses.[13] Amazon (as well as Netflix to an extent) was able to leverage revenue, brand recognition, and infrastructure from its online retail enterprise to also stake a claim. Of course, Hulu is often mentioned alongside these portals, but Hulu was simply a combined enterprise of the owners of NBC, Fox, and ABC and also differentiated by its reliance on advertiser support until its launch of a purely subscriber-funded service in 2015.

The other companies that own portals launched during 2015 and 2016 overwhelmingly own much, if not all, of the content distributed by the portal. The marketplace of internet-distributed portals is quickly becoming so robust—there were nearly one hundred by the end of 2015—as to make launching without a deep library of intellectual property nearly impossible.[14] Part of the barrier to entry that advantages studio portals results from newcomers' lack of an existing library upon which the risk of new series can be balanced. Media economists explain a key strategy of the publishing model as what Miége terms the "dialectic of the hit and the catalogue."[15] In other words, studios, labels, and other media that sell single goods in direct transactions have used a strategy of intentional overproduction in response to the uncertainty of audience tastes because having a diversity of offerings spreads risk. To date, internet-distributed portals have succeeded by curating a breadth of content—both new and acquired library content—that balances the inevitable failures of a high proportion of new content with established hits.

Of course, the development of studio portals has downsides for conglomerates. Following the elimination of the Fin Syn rules, conglomerates effectively bore the full costs and risk of production in integrating production and distribution—costs that broadcast and cable norms split between the studio (30 percent) and the networks or channel licensing the creation of the series

(70 percent). Some attribute the exponential rise in series budgets in the early 2000s to vertical integration; executives reported that it became more difficult to stop spiraling production costs when producers argued that budget overruns were important for the long-term revenue possibilities for series once networks had a stake in them.[16]

The fact that vertical integration is becoming so deeply ingrained as a core strategy of portals is necessary to consider for a variety of reasons. For one, it indicates the likely continuity of key players of US television regardless of the disruption to practices that the affordances of internet-distributed television introduce. Although the pervasive popular and industrial discourse has been one of the "death of television," it is difficult to find substantial casualties. Second, vertical integration emerges as a key curation strategy at a time when little is known about curation tactics. Other than tailoring libraries relative to niche dynamics, the other key explanation of what can be found in a portal's library is that the portal owns the intellectual property.

The prevalence and entrenchment of vertical integration of content and distribution raises concerns given the history of media and its defiance of what Tim Wu describes as the value of a "Separations Principle" for the information economy to which these portals clearly belong.[17] Wu calls for a constitutional approach to the information economy that divides content and distribution companies. Admittedly, the key distribution companies—per Wu's analysis—are the internet service providers. With the exception of Comcast, these companies do not also own stakes in content and are not yet participating in the launch of portals.[18] Nevertheless, it is important to consider the extent to which the concerns that led Wu to advocate for a Separations Principle extend to this context of vertical integration between portals and their intellectual property.

Finally, the effect of vertical integration upon creatives who make this content must be considered. The nature of producing content—the basic business practices involved in production—remains largely unchanged by shifts in distribution technology. But, as established, the strategies of intermediaries such as portals are different than those central to linear distribution technologies. Moreover, the adoption of a subscriber-funded revenue stream—used minimally in broadcast and cable distribution—introduces further change. At a technological level, internet distribution makes possible much more direct relationships between creatives and viewers. As the expansive realm of user-generated, internet-distributed video suggests, intermediary gatekeeping entities that have tremendous power in broadcast and cable distribution can be eliminated. Of course, the last decade has also made clear that intermediaries that help connect viewers to content of interest provide considerable value to both content makers and viewers.

At one level, the vertical integration emerging as characteristic of studio portals connects content makers more directly with audiences in a way that can reduce practices that commonly undermined creatives' autonomy in broadcast- and cable-distributed television. Producing for a studio portal changes the business of production in a number of ways—mostly related to removing the middleman of the channel as scheduler—and affords entities that create content more control. In the network era of US commercial television, series were at the center of multiple, complicated exchanges, creating what economists describe as a dual-product market. Studios created series and first "sold"—although actually rented—them to networks and channels that followed the logics of flow industries and created schedules of programming that attracted audiences that they sold to advertisers. The first market buyer, the network, assessed the value of a show on the basis of its ability to attract the attention of advertiser-desired audiences within a linear schedule. Although this was a phenomenally lucrative exchange for decades, the arrangement is economically messy and inefficient compared with the strategies and practices enabled by nonlinear access and subscriber funding.

Producing original series for a subscriber-funded, studio portal allows those creating programming to transact directly with viewers—or at least with much less intermediation. Much of the arguable "efficiency" of studio portals derives from how they deviate from linear television norms to ensure a single master with a single interest. In the era of the Fin Syn rules, creatives received notes from the studio producing their series as well as from the network licensing the initial run, leading, at times, to contradictory and misaligned creative aspiration. The studio sought to nurture programs in ways that would make them more likely to have the extended production duration valued in syndication sales and other secondary markets, while the network focused on a show's ability to attract an audience in its time slot. Sometimes, these priorities overlapped, but they also regularly diverged. Series produced for studio portals are meant for that portal's library; priorities are not split between competing first and second market dynamics. Moreover, the subscriber-funding model allows emphasis on creating content viewers want without regard to how advertisers might also assess it.

One of the oft-repeated frustrations of studios and creators producing television series for the linear, schedule-defined television environment was their inability to control the time slot of a series; when a show was scheduled could be determinant of success regardless of its content. The success of shows also depended heavily on the extent of network promotion, another factor studios could not control.[19] Studio portals can still promote content differentially—although they have little motivation to deliberately disadvantage any content. The differential promotion of portals can also be far more strategic as comments by Sarandos in Chapter One suggest. In a linear context, differential promotion amounted to offering some programs better time slots and larger promotion budgets than others. In the case of portals, differentiation comes through the ability to target particular shows to particular audiences based on the data known about subscribers' tastes. In practice, this is evident in the way Netflix pushes different shows to viewers in opening pages and recommendations.

Linear networks and channels determine when to continue or cancel series as part of the business of constructing a schedule and managing its inherent capacity constraints. The lack of such constraints for portal library curation largely eliminates the zero-sum situation networks and channels faced that could subordinate creative priorities to the demands of the schedule. Portals do not require content with specific length or a particular number of episodes per season.

In the most radical case to date illustrating the creative possibilities enabled by internet distribution, US comedian Louis C.K. self-funded, produced, and distributed the series *Horace and Pete* directly to audiences who paid a transaction fee per episode or for the season of episodes. *Horace and Pete* was produced without a "channel" or a "studio." That an individual could produce and distribute television in this way tremendously changes the nature of television although this is likely an outlying case. Of course, Louis C.K. is not just any individual, and direct-to-consumer experiments—even by legacy studios—have not been embraced. At this point, it is important to acknowledge this technological possibility although many other industrial aspects may prevent it from ever being widely used.

From the perspective of the business interests of a media conglomerate, studio portals help maximize the value of intellectual property holdings by enabling a direct-to-consumer outlet that eliminates the stake external distributors receive, which is typically 20 to 40 percent of a retail transaction. In launching a studio portal such as CBS All Access, CBS bets that the cost of the portal is recouped in the difference between the revenue it will earn by self-distributing its content instead of licensing it to a portal such as Netflix.[20] Studios have been frustrated by the asymmetrical relationship in which the portals have extensive data about viewing behavior that allow them to far more accurately value content. Now that a substantial portion of the audience has been acculturated to the protocols of internet-distributed video, studio portals provide flexibility and control for intellectual property rights holders at a time of radical change in the business of television. Moreover, they afford the legacy rights holders the depth of viewing behavior data that their internet, upstart competitors have enjoyed.

What remains most uncertain is how well the skills, strategies, and experience of programming a channel will transfer to curating a portal for conglomerates that seek to build studio portals. The greatest uncertainty comes from the lack of knowledge about how building and maintaining a portal differs from deeply entrenched strategies for channels. Strategies related to timeliness and immediacy must be reconsidered and new metrics of success developed. Again, viewers' expectations of content they pay for differ in ways that have meaningful implications for creative priorities.

To be clear, internet distribution via portals—studio or otherwise—does not eliminate the commercial imperatives that have challenged creatives, circumscribed experimentation, and encouraged the persistence of well-known formats and strategies in the television industry. The opportunity of studio portals extends many of the strategic pivots outlined at the beginning of the chapter that shifted the business of television before internet-distributed television emerged. It is too soon to make evidence-based claims about the consequences of studio portals for creativity. Although portals may diminish or eliminate practices that have frustrated creatives producing for broadcast and cable, they will likely create new practices that similarly challenge creatives. As in the case of the yet uncertain implications of subscriber-funded portals, coming critical analysis must understand the practical divergences of studio portals and weigh their limitations against their opportunities.

Timeliness, which so defined linear television, has not been a structuring imperative for many media, and it is the competitive strategies of such media that offer insight into the future of television. Other media that function within the logics of publishing industries such as books, recorded music, and videogames all feature far more asynchronous engagement and thus illustrate possible strategies for internet-delivered television. Of course, time matters to these industries; the moment of initial release of content is inevitably a peak moment of consumption, especially for goods with known properties and industries that push audiences to consume in the first window of availability. But because these internet-distributed media do not have capacity constraints such as a channel schedule, matters of time are not embedded so deeply into their logics.

By way of closing this *preliminary* discussion of *emergent* logics of a *new* distribution mechanism for an *old* medium, I provocatively question how television might operate if it were to follow the conventions of a natively nonlinear medium such as the book industry. Again, this is not a thought experiment relevant to all forms of television, but it is one that has significant potential for reconceptualizing competitive practices and the industrial logics of scripted series.

CONCLUSION: LOOKING OUTSIDE TELEVISION

Ironically, the quintessential analog era publishing industry—the book publishing industry—offers provocative guidance at this moment of reimagining television through internet distribution. In addition to examining strategies emerging in the television industry, it can also be helpful to look at the strategies that developed in other industries that have seemed far removed from television, but now have surprising commonality due to shifts in the models of cultural production that explain television. Such consideration notably moves away from portals, subscriber models, and vertical integration and focuses on how people watch television.

In terms of time of engagement required and the expenditure of leisure, a season of television is most comparable to the time required to read a novel. The book publishing industry is accustomed to the "unscheduled" distribution environment television now faces, as every new book has always competed for the time and attention of readers against other new books, as well as every book ever written. Publishers develop books without the thought of time constraints or strategies related to constructing a linear schedule. They have built their businesses on understanding the rhythms of consumers' leisure needs, independent of underperforming time slots or coordinating the right "lead in."

Far more books are published each year than critics or readers can read so that a "surplus" of books has existed for decades, if not centuries. Concern about "peak TV" or "too much TV" that emerged in 2015 derived from linear distribution norms of synchronous viewing. Indeed, for many viewers, there may have been more hours of compelling television available than most had leisure to view on a week-to-week basis. But the real determinant of "too much" is whether business models and revenue streams can afford this scale of production. Nonlinear distribution does not reveal "success" with the immediacy of media distributed through structures that demand timeliness. Although networks knew the success of their Thursday night schedule by Friday morning, the success of a series for a sub-scriber-funded portal may require months, and even years, to become clear.

As an illustration of the workings of media industries not beholden to timeliness, consider that at any given time a commuter train will be full of riders reading a current release, last year's best sellers, and decades-old classic titles, just as the television viewing in a post-network neighborhood might be similarly dispersed. Book publishers consequently have business models based on creating and circulating content that balances revenue from new titles (new series), new content from known authors (new seasons of established series), and revenue from a backlist (library rights) that account for the asynchronous consumption surplus and non-linear distribution encourage.

As unlikely as it may seem, the book industry, with its publishing model logics, can offer considerable insight at this moment of profound transition in the business of television—particularly for scripted television series. The most useful offering of the book industry may not be in precise practices but in suggesting an alternative paradigm. One of the greatest challenges to rethinking television at this moment of great disruption is being able to imagine television businesses in ways untethered to the logics of television's past. To paraphrase digital scholar Nicholas Negroponte, perhaps the future of television is thinking about television series as we have novels.[1]

The book industry offers a long history of its medium's adaptation to shifting distribution forms (hardback, paperback, e-book) through corresponding adjustments in business models (subscription and circulating libraries, direct sale) and the establishment of sectors with distinct logics (mass market, award contender, specialized topics, academic) that aid thinking about the business of television in the different ways required in an era in which television is free from linear schedules.

The book industry has endured considerable disruption of its own in the last few decades although curiously less instigated by digitization of media and internet distribution than other media industries. The economics of book publishing shifted as the value of the paperback market waned after retail superstores emerged and heavily discounted hardbacks; conglomeration swept through the industry leaving a handful of global giants; and the rise and fall of bricks and mortar retail giants altered distribution just before online retailers and e-books disrupted norms yet again.[2]

And this provocation can be turned back on itself. Internet-distributed television has been most successful when offering viewers access to a library of content for a fee, what I have argued as characteristic of a subscriber model—a model very different than used by the contemporary book industry. This may be the logic of a preliminary era of internet distribution—just as circulating libraries developed in the eighteenth century before public libraries and affordable book pricing—but its deviation from strategies common in internet-distributed sectors of print and audio industries is notable. Likewise, although the print industries of magazines and newspapers and the recording industry preceded the television industry in adapting practices for the affordances of internet distribution, these industries have most minimally used vertical integration as a strategy for these revised conditions. Of course, there are peculiar characteristics of the production and consumption of different media that may explain the discrepant strategies, or there may be valuable lessons at hand.

Internet distribution and its affordances that allow for some sectors of television to operate consistently with the publishing model—or require a subscriber model—remain too nascent to derive certain implications for creativity or social and political impact. As adjustments introduced by nonlinear access continue to emerge in coming years, critics can expect to revisit or relinquish many theories about television and its operation in society. More precisely, much reassessment will be needed: established theories may continue to hold for sectors that remain governed by flow logics, but new theories will be needed for those sectors of television more consistent with publishing and subscriber models. Just as the advertising industry struggles to determine new metrics and practices, so too must critical analysis refocus and re-theorize. Importantly, internet-distributed television is not fundamentally unlike everything that has come before.

The profound change that has only just begun requires considerable creation of new understandings before returning to the evaluations at the heart of efforts to understand the role of media in society. The logics of the television industries continue to shift. Strategies of the moment can be discerned, but remain fleeting. In time, new norms will emerge. As these industries evolve, we can continue to identify practices and strategies, consider their consequences, and begin to collect evidence that enables theorization of implications. Rather than also remaining blinded to new possibilities by expectations of the continuation of a broadcast paradigm, we can begin applying what we have known of other media industries as we make sense of television's continued evolution.

Introduction

1. Amanda D. Lotz, *The Television Will Be Revolutionized*, 2nd rev. ed. (New York: New York University Press, 2014).
2. Henry Jenkins, *Convergence Culture: Where Old and New Media Collide* (New York: New York University Press, 2005), 3.
3. As Finneman explains, "the fundamental architecture" of digital media can be "modified, changed or suspended by means of individual messages sent in the very same medium." Niels Ole Finneman, "Public Space and the Coevolution of Digital and Digitized Media," *MedieKultur: Journal of Media and Communication Research* 22, no. 40 (2006), doi:10.7146/mediekultur. v22i40.1318.
4. Lynn Spigel, "Introduction," in *Television after TV: Essays on a Medium in Transition*, ed. Lynn Spigel and Jan Olsson (Durham: Duke University Press, 2004), 2.
5. Raymond Williams, *Marxism and Literature* (Glasgow: Oxford University Press, 1977); Roger Silverstone, *Television and Everyday Life* (London: Routledge, 1994); John Corner, *Critical Ideas in Television Studies* (Oxford: Oxford University Press, 1999).
6. Lotz, *The Television*; Michael Curtin, "On Edge: Cultural Industries in the New-Network Era," in *Making and Selling Culture*, eds. Richard Ohmann, Gage Averill, Michael Curtin, David Shumway, and Elizabeth Traube (Hanover, NH: Wesleyan University Press, 1996), 181–202; Mark C. Rogers,

Michael Epstein, and Jimmie L. Reeves, "*The Sopranos* as HBO Brand Equity: The Art of Commerce in the Age of Digital Reproduction," in *This Thing of Ours: Investigating the Sopranos*, ed. David Lavery (London: Wallflower Press, 2002), 42–57.

7. Jenkins, *Convergence Culture.*

8. Jenkins, *Convergence Culture*, 13–14; Lisa Gitelman, *Always Already New: Media, History and the Data of Culture* (Cambridge: MIT Press, 2008).

9. Gitelman, *Always Already New.*

10. "Cable" is defined here more culturally than technologically. Admittedly, digital cable technologies have a wider range of affordances than broadcast and their analog predecessor, however, even by 2017, "cable" predominantly remained a one-to-many service.

11. John B. Thompson, *Merchants of Culture: The Publishing Business in the Twenty-first Century*, 2nd ed. (New York: Plume, 2012), 11.

12. See Van Esler for helpful detailed descriptions of the basic business of Hulu, Netflix, YouTube and MVPD-delivered on demand services. Mike Van Esler, "Not Yet the Post-TV Era: Network and MVPD Adaptations to Emergent Distribution Technologies," *Media and Communication* 4, no. 3 (2016): 131–41.

13. Admittedly, the bigger point of confusion here is in understandings of "cable" distribution. At the risk of excessive detail, the transition to the DOCSIS (data over cable service interface specification) standard in the US allowed cable providers the affordance of nonlinear distribution. Although this technological capacity exists and is used in "on-demand" service, "cable"—as of 2017—remains more defined by linear distribution of content and related practices. With the exception of Comcast making library content it owns available, the cable services do not curate a library of video on demand offerings. At some point in the future, the distinction between cable- and internet-distributed television may become negligible due to cable's adoption of different industrial practices and related protocols.

14. Web TV offered a product and service that enabled clients to use television sets to access the web. Microsoft acquired it in 1997, and it was shut down in 2013.

15. Sherman, Waterman, and Jeon note that none of the significant players in the current market (2014) were present before 2005. Ryland Sherman, David Waterman, and Yongwoog Jeon, "The Future of Online Video: An Economic Perspective." Paper Presented at the Future of Broadband Regulation Workshop, Federal Communication Commission, Washington, DC, May 29–30, 2014.

16. Diane Mermigas, "IP Promises Video-to-go as Next Big Media Wave," *Hollywood Reporter*, October 11/17, 2005, 8, 76.

17. Amanda Lotz, "How OTT Hides Television's Revolution," *Broadcasting & Cable*, March 8, 2016, http://www.broadcasting cable.com/blog/bc-beat/guest-blog-how-ott-hides-television-s-revolution/154442.

18. As I argue in Lotz, 2016, "Others have termed these as platforms or apps; all of these terms carry with them wide-ranging and complicated uses. Portal first emerged in relation to computing to describe a website or service that provided internet service, then to websites or services that provided access to a variety of content, for example, AOL as an internet portal (*Oxford English Dictionary*). Portal has otherwise fallen from use, freeing it for this use that draws on the notion of a gateway to video content. Platform continues to be used extensively as computing systems (Ian Bogost and Nick Montfort, "Platform Studies: Frequently Asked Questions," *Digital Arts and Culture,* 2009, http://nickm.com/if/bogost_montfort_dac_2009.pdf) or as a set of interfaces and tools (Facebook platform), while apps describe too broad an array of programs or software. Platform and app can also connote viewing on particular devices and this theorization of portal seeks to be device agnostic. The entities distinguished as portals here provide access to a wide range of content and the thirteenth-century definition of the term emphasizes their function as a gateway, in this case to a library of television programs." See Amanda D.

Lotz, "The Paradigmatic Evolution of U.S. Television and the Emergence of Internet-Distributed Television" *Icono 14 Journal of Communication and Emergent Technologies* 14, no. 2 (2016): 122–42, http://www.icono14.net/ojs/index.php/icono14/article/view/993/566.

19. Stuart Cunningham and David Craig, "Online Entertainment: A New Wave of Media Globalization?," *International Journal of Communication* 10 (2016). Notably, "YouTube" varies considerably in different contexts. In the US, it has a minimal role in long-form content distribution although this is a more central function in other national contexts.

20. Louis C.K.'s *Horace and Pete* is a notable exception and certainly not a large-scale operation.

21. Notably BBC's iPlayer provides many of these capabilities as early as 2007.

22. Daniel Frankel, "Cable's Ratings Collapse," *Fierce Cable*, March 12, 2015; Cynthia Littleton, "TV's New Math: Networks Crunch Their Own Ratings to Track Multiplatform Viewing," *Variety*, February 11, 2015, http://variety.com/2015/tv/features/broadcast-nets-move-closer-to-developing-ratings-that-con sider-auds-delayed-viewing-habits-1201430321.

23. Josef Adalian and Leslie Shapiro, "The 2015–16 TV Season in One Really Depressing Chart," *Vulture*, March 16, 2016, http://www.vulture.com/2016/03/2015–2016-tv-season-in-one-de pressing-chart.html.

24. Suzanne Vranica and Joe Flint, "Digital Takes Its Toll on TV's Upfront Ad Sales," *Wall Street Journal*, March 26, 2015, B1.

25. Also, for much of their history, these services were linked closely with the theatrical films that provided most of their programming. This too led them and their revenue model to go unconsidered.

26. Shalini Ramachandran, "Niche Sites Like Zombie Go Boom Target Underserved Markets," *Wall Street Journal*, April 12, 2016, http://www.wsj.com/video/niche-sites-like-zombie-go-boom-target-underserved-markets/B57A7513–84B8–456A-B3B0–6F934225C6FA.html.

Chapter 1

1. As opposed to building business models that would allow the original licensing of a show to adequately compensate its creation—more the case in formats such as reality television and news and sports content that is not typically sold and resold over an extensive period of time.

2. A variety of revenue models have emerged to support different internet-distributed television services. Examples of advertiser, subscriber, public service, and transaction payment can be identified in support of this nascent stage of internet-distributed television, although as of 2017, subscriber funding was the dominant revenue mechanism. Of course, other revenue strategies were possible for broadcast and cable distribution technologies so that distribution technology does not innately determine this protocol. The affordances of distribution technologies do encourage and discourage various revenue models.

   Revenue models delineate many other aspects of industrial operation and viewer experience so extensively that it is necessary to separately explore the emerging sectors of advertiser- and subscriber-supported internet-distributed television. In particular, curation strategies derive from whether a portal measures success by paying subscribers or unique viewers.

3. Gillian Doyle, "Resistance of Channels: Television Distribution in the Multiplatform Era," *Telematics and Informatics* 33 (2016): 693–702.

4. Williams, *Marxism and Literature*.

5. William Uricchio, "Contextualizing the Broadcast Era: Nation, Commerce, and Constraint," *Annals of the American Academy of Political and Social Science* 625 (2009): 60–73. The long-term issue is a question of whether broadcast and cable distribution have affordances that internet distribution lacks. In 2017, there are reasonable questions about the robustness of internet distribution and its infrastructure in terms of whether it can bear the entirety of contemporary video distribution. Such capacity may not be currently feasible, but will undoubtedly develop in time. It is at that point that the pluriformity may decrease.

6. Bernard Miége, *The Capitalization of Cultural Production* (New York: International General, 1989), 145–46.

7. Miége, *The Capitalization*, 133. Within this school of media industry analysis, Miége argues scholars might either "study the strategies of the concerned social and economic actors in order to identify their components and clarify their interactions or contradictions, with the hope that from this patient observation we may ultimately obtain some general information that will renew our scientific knowledge," or "start with the knowledge already available and take the risk of formulating hypotheses that clarify and complete what we already know about the cultural industries, and then, with the help of this new 'theoretical grid', attempt to determine, beyond the evolution of strategies, the significance of current tendencies" (133).

8. Jean-Guy Lacroix and Gaëtan Tremblay, "The Emergence of Cultural Industries into the Foreground of Industrialization and Commodification: Elements of Context," *Current Sociology* 45 (1997): 11–37, 43.

9. Patrice Flichy, "Industries Culturelles," in *Le Dictionnaire de la Communications*, tome 2, ed. L. Sfez (Paris: Presses Universitaires de France, 1993).

   A note on terminology: Given the reliance on translation, it is difficult to be certain of the precision of use of logics versus models in the original text. The translation is imprecise and uses the terms interchangeably. My use is consequently not entirely consistent with Miége and Flichy, but very much inspired by them. Here, models are large-scale organizing categories—written press, flow, publishing/commodity, and now subscription, whereas logics are industrial practices that emerge within models, as suggested by Thompson.

10. Lacroix and Tremblay, "The Emergence."

11. Miége, *The Capitalization*, 145–46.

12. Miége, *The Capitalization*, 144–45.

13. The other commonly transacted medium is recorded music. This case is less informative because of the substantially

greater regard with which repeat listening of music is valued (listeners want to hear a song or album they buy again and again), whereas few books are reread and few television series are rewatched.

14. At least this is the case in the United States. Television on DVD has persisted when series are not otherwise available.

15. Computed from totaling 2,495.97 electronic sell-thru, 5,321.63 for subscription video on demand, and 3,149.38 pay per view (2,495.97/10,966.90 = .2276) Figures from Statista, "Subscription video on demand market in the U.S.," 2015; http://www.statista.com/statistics/456796/digital-video-revenue-type-digital-market-outlook-worldwide/.

16. David S. Evans and Michael A. Salinger, "Why Do Firms Bundle and Tie: Evidence from Competitive Markets and Implications for Tying Law," *Yale Journal on Regulation* 22 (2005): 37.

17. Gregory Crawford and Joseph Cullen, "Bundling, Product Choice, and Efficiency: Should Cable Television Networks Be Offered á la Carte?" *Information Economics and Policy* 19, no. 3–4 (2007): 379–404.

18. Lacroix and Tremblay, "The Emergence."

19. Yannis Bakos and Erik Brynjolfsson, "Bundling and Competition on the Internet," *Marketing Science* 19, no. 1 (2000): 63–82.

20. Of course, Nickelodeon too is subscriber funded in part. However, having any significant component of advertiser funding tends to make that the structuring logic, and the inability to specifically subscribe to Nickelodeon—rather than having it included in a subscription to a larger, provider-determined bundle of channels, further distinguishes it from Noggin's exclusively subscriber-funded model.

21. Corey Allan, Arthur Grimes, and Suzi Kerr, *Value and Culture: An Economic Framework* (Wellington: Manatu Taonga-New Zealand Ministry for Culture and Heritage, 2013).

22. Allan, Grimes, and Kerr, *Value and Culture*, 8.

23. A site such a YouTube illustrates this strategy in the extreme. By not paying for content, it is able to offer an exceptionally vast library.

24. At this point, this is more clearly seen in music distribution portals. As Wikstrom identifies, in music there is a history of an ownership model that has never really existed for television. Also, music tends to be replayed much more than television is rewatched, so the specificity of medium also distinguishes strategies in the emergent digital library industries. Patrik Wikstrom, "A Typology of Music Distribution Models," *International Journal of Music Business Research* 1, no. 1 (2012): 7–20.

25. Some linear channels have attempted this—for example, Cartoon Network services a youth audience by day, but programs Adult Swim in its late night hours. In another case, the Spike channel emphasized its traditional masculinity brand during prime time although many of its daytime hours targeted a broader audience with off-net episodes of *CSI*.

26. Reported by Carolyn Newman of Amazon Studios to Charlotte Howell, telephone interview, January 13, 2015. Published in Charlotte Howell, "Divine Programming: Religion and Prime-Time American Television in the Post-Network Era." PhD diss., The University of Texas at Austin, 2016.

27. Alan Sepinwall, "Ted Talk: State of the Netflix Union Discussion with Chief Content Officer Ted Sarandos," *HitFix*, January 26, 2016, http://www.hitfix.com/whats-alan-watching/ted-talk-state-of-the-netflix-union-discussion-with-chief-content-officer-ted-sarandos.

28. Sean Ludwig, "Netflix Says More People Watched *Hemlock Grove* on First Weekend than *House of Cards*," April 22, 2013, http://venturebeat.com/2013/04/22/netflix-hemlock-grove-first-weekend/.

29. *The Hollywood Reporter*, Full TV Executive Roundtable, August 10, 2016, http://www.hollywoodreporter.com/video/watch-thr-s-full-tv-918281.

30. Ted Striphas, "Algorithmic Culture," *European Journal of Cultural Studies* 18, no. 4 (2015): 395–412.
31. Jeremy Wade Morris, "Curation by Code: Infomediaries and the Data of Mining Taste," *European Journal of Cultural Studies* 18, no. 4–5 (2015): 446–63.
32. Michael D. Smith and Rahul Telang, *Streaming, Sharing, Stealing: Big Data and the Future of Entertainment* (Cambridge: MIT Press, 2016), 75.
33. Statista, "Leading Reasons Why Netflix Subscribers in the U.S. Subscribed to Netflix as of January 2015"; data reported in eMarketer from a Cowan & Company study, methodology not specified; http://www.statista.com/statistics/459906/reasons-subscribe-netflix-usa/.
34. Richard E. Caves, *Creative Industries: Contracts Between Art and Commerce* (Cambridge: Harvard University Press, 2000), 8–9.

Chapter 2

1. Richard Roehl and Hal R. Varian, "Circulating Libraries and Video Rental Stores," *First Monday* 6, no. 5–7 (2001), http://pear.accc.uic.edu/ojs/index.php/fm/article/view/854/763.
2. The circulating libraries are the stronger predecessor because the video rental business typically used a per rental fee rather than a regular fee for unlimited access to a catalog of holdings.
3. Lacroix and Tremblay, "The Emergence," 63.
4. Amanda D. Lotz, "If It's Not TV, What Is It? The Case of U.S. Subscription Television," in *Cable Visions: Television beyond Broadcasting*, ed. Sarah Banet-Weiser, Cynthia Chris, and Anthony Freitas (New York: New York University Press, 2005).
5. Bakos and Brynjolfsson, "Bundling and Competition on the Internet."
6. Smith and Telang, *Streaming, Sharing, Stealing*.
7. Notably, that "functionally zero" cost results from the current way contracts are structured. Different norms—such as

licensing costs based to actual viewership—are feasible and could create marginal costs.

8. Valerie-Anne Bleyen and Leo van Hove, "To Bundle or Not to Bundle? How Western European Newspapers Package Their Online Content," *Journal of Media Economics* 23 (2010): 117–42.

9. R. Venkatesh and R. Chatterjee, "Bundling, Unbundling, and Pricing of Multiform Productions: The Case of Magazine Content," *Journal of Interactive Marketing* 20.2 (2006): 21–40.

10. Crawford and Cullen, "Bundling, Product Choice."

11. See Doyle for analysis of digitization's implications for television windowing practices. Gillian Doyle, "Television Production, Funding Models, and Exploitation of Content," *Icono 14 Journal of Communication and Emergent Technologies* 14, no. 2 (2016): 75–96; Gillian Doyle, "Digitization and Changing Windowing Strategies in the Television Industry: Negotiating New Windows on the World," *Television and New Media* 17, no. 7 (2016): 1–17. Also Ronen Shay, "Windowed Distribution Strategies for Substitutive Television Content: An Audience-Centric Typology," *International Journal on Media Management* 17 (2015): 175–93.

12. Bruce Owen and Steve Wildman, *Video Economics* (Cambridge: Harvard University Press, 1992).

13. Of course, the technology often reasserts geography to maintain business strategies through geofiltering that disallows access if using an IP address outside of the licensed region.

14. Bakos and Brynjolfsson, "Bundling and Competition on the Internet."

15. Notably, this is changing and many providers have established monthly caps of one terabyte. This cap—estimated to allow for 700 hours of HD video screening—should not produce negative effects for most who rely on home internet for portal streaming at the current standard.

16. A recent deal between Netflix and Comcast suggests some adjustment. Comcast will include access to Netflix in its X1

user interface. Details of the arrangement have not been made public, but it is likely that Comcast will receive some compensation for those who sign up using the interface, but will not receive nearly the even split of the linear service.

17. A 2016 article reports that Apple's initial share of fees was 30 percent but would drop to 15 percent. Ben Munson, "Apple Will Reportedly Take Smaller Cut of VOD Subscriptions Sold in App Store," *Fierce Cable*, November 17, 2016, http://www.fierce cable.com/broadcasting/apple-will-reportedly-take-smaller-cut-vod-subscriptions-sold-app-store.

18. Liam Boluk/Matthew Ball, "The State and Future of Netflix v. HBO in 2015," *ReDef*, March 5, 2015, http://redef.com/original/the-state-and-future-of-netflix-v-hbo-in-2015.

19. Bleyen and van Hove, "To Bundle," 119.

20. In May 2016, Netflix announced a licensing deal with Univision to provide a second window for a few of its original series, and HBO licenses its content in international markets where the revenue is greater than self-distribution and also licenses library content to Amazon Video. To date, both HBO and Netflix have made limited use of such deals and typically not for the content in highest demand.

21. Matthew Ball, "Letting It Go: The End of Windowing (and What Comes Next)?" *ReDef*, August 26, 2016, http://redef.com/original/letting-it-go-the-end-of-windows-and-what-comes-next?

22. Bill McConnell, "Never Say Never," *Broadcasting & Cable*, September 13, 2004, 1, 10.

23. The US PBS also makes a lot of content available via website. The ongoing underfunding of this outlet makes it difficult to expect more.

24. For specific examples, see the interviews in the Creatives section of Michael Curtin, Jennifer Holt, and Kevin Sanson, eds., *Distribution Revolution: Conversations about the Digital Future of Film and Television* (Oakland: University of California Press, 2014).

25. Joe Adalian and Maria Elena Fernandez, "The Business of Too Much TV," *Vulture*, May 2016, http://www.vulture.com/2016/05/peak-tv-business-c-v-r.html.

26. Maureen Ryan, "Netflix, Binging and Quality Control in the Age of Peak TV," *The Huffington Post*, August 27, 2015, http://www.huffingtonpost.com/entry/netflix-binging-and-quality-control_us_55df5816e4b029b3f1b1f625.

27. As of June 2016, half of US homes accessed a subscriber-funded service. Nielsen Media Research, "Milestone Marker: SVOD and DVR Penetration Are Now on Par with One Another," *Nielsen Insights*, June 27, 2016, http://www.nielsen.com/us/en/insights/news/2016/milestone-marker-svod-and-dvr-penetration-on-par-with-one-another.html.

28. Joseph Turow, *Breaking Up America: Advertisers and New Media World* (Chicago: University of Chicago Press, 1998).

29. Circulating and subscription libraries were distinct entities although both had a similar economic transaction. Subscription libraries tended to collect scholarly works and made them available to a narrow elite of subscribers while circulating libraries gained popularity for making novels (a delegitimized form of culture at the time) available to the general populace. The cases of subscriber-funded portals are more like circulating libraries, but that can be confusing since the distinction between circulating and subscription libraries is not widely known.

Chapter 3

1. Miége, *The Capitalization*; Lotz, *The Television*.

2. Jennifer Holt, *Empires of Entertainment: Media Industries and the Politics of Deregulation, 1980–1996* (New Brunswick: Rutgers University Press, 2011), 54–66.

3. Lotz, *The Television*, 94.

4. Amanda D. Lotz, *The Cable Revolution* (in press).

5. Interview with John Landgraf, *KCRW's The Business*, September 18, 2015.

6. Lotz, *The Cable Revolution*.

7. Notably, satellite and telcos—new video service competitors that historically provided phone service such as Verizon and AT&T—had already been paying broadcasters. As later entrants to competition, they paid fees twenty to fifty percent higher than incumbent cable distributors. Cynthia Littleton, "Pay TV under Pressure," *Variety* Thought Leader Report, July 2016, 4.

8. Cited in Littleton, "Pay TV under Pressure," 3.

9. S.N.L. Kagan, "Broadcast Retransmission Fees in the U.S. from 2006–2021," Statista, https://www-statista-com.proxy.lib.umich.edu/statistics/256358/broadcast-retransmission-fees-in-the-us/.

10. Laura Martin and Dan Medina, "The Future of TV," *Needham Insights*, July 11, 2013.

11. Todd Spangler, "Pay-TV Prices Are at the Breaking Point—And They're Only Going to Get Worse," *Variety*, November 29, 2013, http://variety.com/2013/biz/news/pay-tv-prices-are-at-the-breaking-point-and-theyre-only-going-to-get-worse-1200886691/.

12. Littleton, "Pay TV under Pressure," 8.

13. For a more detailed look at Netflix role in U.S. internet-distributed television see Lotz, *The Cable Revolution*.

14. Shalini Ramachandran, "Niche Sites Like Zombie Go Boom Target Underserved Markets," *Wall Street Journal*, April 12, 2016, http://www.wsj.com/video/niche-sites-like-zombie-go-boom-target-underserved-markets/B57A7513–84B8–456A-B3B0–6F934225C6FA.html.

15. Miége, *The Capitalization*, 137.

16. Lotz, *The Television*, 94.

17. Tim Wu, *The Master Switch: The Rise and Fall of Information Empires* (New York: Knopf, 2011), 304.

18. As this book went to press, AT&T announced plans to purchase Time Warner, which would likewise allow a major video/internet distributor to own significant content. It was unclear whether the purchase would be allowed by government regulators.

19. The initial post Fin Syn vertical integration has been uniformly viewed as a negative development for creative and storytelling diversity. Indeed, it is clearly the case that it resulted in the demise of independent studios that prospered and produced much-heralded programming in the Fin Syn era. The vertical integration at the heart of studio portal is of a somewhat different character. First, the competitive terrain has changed extensively from the monopsony days of three or four buyers of television series. Second, the competitive field now consists of several such vertically integrated entities. It is also true that these vertically integrated media companies can be seen as erecting even greater barriers to entry for new competitors because of the need for new competitors to have production and distribution facilities, and arguably a library of existing intellectual property as well.

20. See analysis by Dan Schechter, "Why the Streaming TV Boom Is about More than Just Netflix," *The Wrap*, July 27, 2016, http://www.thewrap.com/streaming-tv-boom-ott-more-than-just-netflix-guest-blog/.

Conclusion

1. Nicholas Negroponte, *Being Digital* (New York: Vintage, 1995), 48.
2. Thompson, *Merchants of Culture*, 22.

INDEX

A lá carte, 17, 58

Advertiser support, 13, 16, 19, 23, 25, 34, 52, 55, 59

Advertiser-supported video on demand (AVOD), 7

Affordances, 1–2, 4–6, 11, 13, 16–20, 27, 29, 33–5, 42–3, 57, 61, 68, 72, 81

Algorithms, 28, 45, 50

Amazon Video, 6, 26, 64

Bakos, Yannis and Brynjolfsson, Erik, 22, 37

Ball, Matthew, 55

BamTech, 45

Branding, 24, 26

Broadcasting
distribution, 2–6, 10–12, 18–20, 23–4, 33, 41, 52, 65–7, 70, 73, 82
sectors, 62–3

Bundling, 22, 37–8, 48, 58, 68

Cable
channels, 4, 6, 8–9, 13, 23–6, 28, 34, 38, 40, 58, 63–73, 75–7

on demand, 2–3, 6–7, 34, 36
sectors, 63–4
service providers, 2, 7, 67–70

Caves, Richard, 38

CBS All Access, 8, 29, 47, 75

C. K., Louis, 21, 75

Circulating libraries, 16, 34, 81

Club logic, 18, 22, 35

Conglomerated niche, 26–7, 58

Corner, John, 3

Cunningham, Stuart and Craig, David, 10

Curation, 8, 22–4, 27, 29, 36, 39–40, 43, 46, 48, 72, 75–6

Data, 27–9, 38, 41, 43–7, 56, 74–6

Demographics, 60

Digital video disk (DVD), 10, 15–16, 21, 41, 55, 67

Digital video recorder (DVR), 2, 15

Direct-to-consumer (see also *Horace and Pete*), 4, 48, 74–5

Exclusivity, 16, 25, 28, 48–9, 53

Financial Interest and Syndication Rules, 64

Flichy, Patrice, 18, 36

Flow model, 17–20, 33, 35

Gitelman, Lisa, 3–4

Hastings, Reed, 27

HBO (other subscriber-funded channels), 13, 34, 36–7, 40, 42, 48, 54, 56, 58
  monthly profit, 55

HBO Now, 6, 8–9, 45, 47–8

Hemlock Grove, 27

Horace and Pete, 21, 75

House of Cards, 27

Hulu, 6–7, 11, 53, 64, 71

Intellectual property, 2, 8, 10, 38, 40, 47, 62, 66–7, 71–2, 75

Internet protocol, 6, 8, 15–16

iTunes, 10, 37, 67

Jenkins, Henry, 3–4

Lacroix, Jean-Guy and Tremblay, Gaëtan, 18, 35–6

Library, 4, 8, 21–2, 24–6, 29–30, 37, 39–41, 44, 47, 58, 71, 75, 81

Licensing, 16, 24, 26, 28–9, 40–5, 48, 62–6, 74–5

Linear television, 3–4, 6, 8–9, 15–6, 22–23, 25–6, 28–30, 22, 36–7, 43, 47, 54, 73–6, 79–81

Media
  definition of, 1–5

Míege, Bernard, 17–20, 35–7, 48–9, 71

Multichannel video programming distributors (MVPDs), 68–70

Nathanson, Michael, 66

Neo-network era, 3

Netflix, 6–9, 11, 26–8, 41, 44, 46–8, 53, 55, 57, 64, 70–1, 74–5
  reasons for subscribing, 30

Net (network) neutrality, 70

Niche, 24–8, 47, 58, 72

Noggin, 24–5, 47

Nonlinear, 4–5, 7–11, 13–31, 57, 61, 80, 82
  definition of, 2

Over the top (OTT), 15–6

Pluriformity, 17

Portals
  definition of, 8

Post-network era, 3, 80

Protocol, 6, 8, 10–13, 18, 20–1, 23, 33, 61, 75
  definition of, 3–4,

Publishing model, 10, 18–23, 31, 37–8, 44, 46, 48, 71, 76

Retail, 10, 25, 37, 71, 75, 81

Sandler, Adam, 27, 44

Sarandos, Ted, 27, 74

Satellite, 36, 47, 58, 62–3, 67, 69–70

Scarcity, 16, 18, 22, 24, 28, 40, 49

Schedule, scheduling, 3–4, 8–9, 12, 14–16, 18–19, 22–3, 25, 28, 33, 35–40, 63–5, 73–5, 79–81

SeeSo, 8, 29, 47

Separations principle, 72

Silverstone, Roger, 3

Skinny bundles (internet-distributed MVPDs), 53

Smith, Michael D. and Telang, Rahul, 28, 38

Spigel, Lynn, 3–4

Studios, 13, 16, 29–30, 44, 54, 63–5, 67–8, 71, 73–6

Studio portals, 14, 29, 63–5, 67–8, 71, 73–6

Subscriber model, 13–14, 18, 39–56, 58–9, 74, 79, 81–2

Subscription libraries (see circulating libraries), 16, 66, 81

Subscription video on demand (SVOD), 7–8

Syndication, 30, 67, 74

Thompson, John B., 5

Transaction video on demand (TVOD), 7

TVIII, 3

Twitch, 22

Uricchio, William, 17

Vertical integration, 13–14, 29, 40, 61–81

Videocassette recorders (VCRs), 2, 15

Video on demand (VOD), 2, 6–7, 34, 36

Viewer interface, 9, 41, 45

Watch ESPN, 22

Web TV, 6–8

Williams, Raymond, 3, 17

Windowing, 16, 40, 48, 55, 77

Written press model, 17–19, 31, 35–6, 38

Wu, Tim, 72

WWE Network, 24, 45, 47

YouTube, 6, 10, 22, 53

CPSIA information can be obtained
at www.ICGtesting.com
Printed in the USA
LVOW10s1255311017
554442LV00004B/649/P